Sustainable Energy

Dai J. Redshaw is Head of Science at Kavanagh College, Dunedin, and Keith R. Dawber lectures in the Physics Department at the University of Otago. Both authors have a long involvement in researching and teaching strategies in the use of renewable energy sources.

Published by
University of Otago Press

Sustainable Energy
Options for New Zealand

Dai Redshaw and Keith Dawber

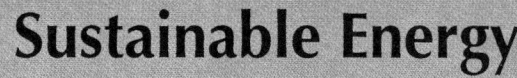

Published by University of Otago Press
56 Union St, Dunedin, New Zealand

© Dai J. Redshaw and Keith R. Dawber 1996
First published 1996
ISBN 0 908569 95 5

Cover design by Marc Mundy
Cover photo of windmills by Denis Pagé
Cover photo of landscape by Neville Peat
Layout by Jenny Cooper
Diagrams by Peter Scott
Printed through Condor Production Ltd, Hong Kong

Contents

Preface

We have written this book in the hope that it will be useful for anyone who wishes to consider New Zealand's future energy options. We have paid particular attention to international concern over global warming, and how New Zealand can meet its obligations under the Climate Convention which was signed by over 160 countries at the Earth Summit in Rio de Janeiro in 1992, and which came into force on 21 March 1994 after ratification by over fifty countries.

We have also written this book for those who wish to examine renewable energy as physics in context. Some of the matters that we discuss are by their very nature matters of opinion – we have been prepared to indicate where our sympathies lie. We have also made some speculative suggestions, which may or may not prove practical. In these cases the opinions given are those of the authors and are not necessarily the policies of the authors' employers or of the Foundation for Research, Science and Technology (FORST).

We have tried to ensure the accuracy of the data which we present. Unfortunately, within the time available to prepare this publication, it has not been possible to consult all the original publications in which this data has been published, but rather we have had to rely on the accuracy of the review articles we have used. When doing calculations we have worked to two significant figure accuracy, unless greater accuracy is important to the calculation. Because of this, there may appear to be discrepancies between data, resulting from the rounding up, or down, of numbers.

We have attempted to avoid technical terms or jargon in our discussions. However, there are sometimes no sensible alternatives. Where terms we believe require definition are introduced, a definition is provided in the glossary on page 105.

As the noun 'milliard' is little used these days, we have used the term 'billion' to signify one thousand million (10^9) and 'trillion' to signify 10^{12}. We have discussed the SI method of dealing with other large numbers in the box on 'Energy and Power' on page 10.

This book resulted from one of the authors (DJR) being granted a FORST Teacher Fellowship, which enabled him to attend the World Renewable Energy Congress III in Reading UK during September 1994, and to spend one term in the Physics Department of the University of Otago working with the other author (KRD), who is a senior lecturer there.

The authors wish to thank the University of Otago and the Board of Trustees of Kavanagh College for their support throughout this work. We also thank the many people who have been prepared to assist us either in discussions, or by drawing our attention to useful information. Particular thanks should go to Elwyn Hughes for his assistance over matters relating to hydro-power, to Paul Mierzejewski on solar power, and to Molly Melhuish for her valuable corrections to the manuscript. Also, we thank Marlyn Jakub for help with the nuclear energy chapter and, of course, our publishers.

DAI REDSHAW & KEITH DAWBER

I thought that it would last my time—
The sense that, beyond the town,
There would always be fields and
farms.

... it isn't going to last,
... before I snuff it, the whole
Boiling lot will be bricked in
... all that remains
For us will be concrete and tyres.

Philip Larkin
Going Going

1 Energy in New Zealand's Future

New Zealand is an economically developed country. Because of this development, we use a large amount of energy. Fig. 1.1 lists the per capita annual energy consumption and the gross national product (GNP) for some countries in 1989. As can be seen, in New Zealand we have about half of the per capita energy consumption and GNP of the United States, but many times more than that of Brazil and Bangladesh.

Developed countries have taken the apparently unlimited supply of energy for granted. Since the industrial revolution, as one energy source began to run out, another has been ready to take its place. Wood fuel was replaced by coal as forests were depleted. Early this century oil became the convenient fuel, and natural gas is now preferred in many situations. Nuclear power, which was developed after World War II, was expected to be so cheap that there would be no point in metering it, but although nuclear reactors supply large quantities of energy in some countries, particularly France and Belgium, it has not fulfilled earlier expectations.

Country	Per capita annual consumption (GJ)	Per capita GNP ($US)
New Zealand	151	11800
United States	295	21100
Australia	211	17080
Brazil	23	2680
Bangladesh	2	200

Figure 1.1 1989 energy consumptions and GNPs.

The trend of cheaper energy was reversed by the 'oil shocks' of 1973 and 1978. This, together with the problems that have beset nuclear energy, have led many people to talk of an 'energy crisis' in the future.

Recently it has been recognised that there is a possibility of global warming caused by the build-up of greenhouse gases, especially carbon dioxide (CO_2), in the atmosphere. If global warming is happening, and the evidence is strengthening that it is, then this will place a limit on the use of fossil fuels well before such fuels run out (Fig. 1.2).

New Zealand has been a pioneer in the development of hydro-electricity and this

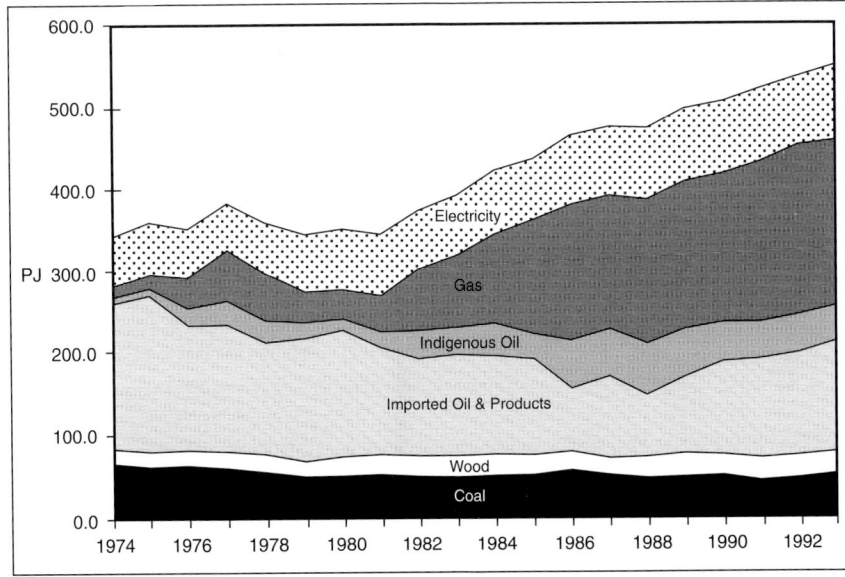

Figure 1.2 Annual New Zealand primary energy supply in petajoules. (Ministry of Commerce)

has allowed us to be less dependent on oil or coal-fired electricity generation than are most countries; we have also avoided the need to construct nuclear power stations. Although there are problems with expanding our hydro-generation, we are in a good position to take advantage of new technologies, especially those which obtain energy from biomass, solar energy and wind. As we hope to show in this book, use of these energy sources could take care of our energy requirements in the foreseeable future.

One characteristic of a developed country is that its citizens are very mobile and use considerable energy for transport. Because of this, we have devoted a chapter (Chapter 12) to the problem of maintaining energy supplies for the transport sector. New Zealand has been innovative in the development of alternative transport fuels. We have converted many of our cars to LPG and have a programme to convert both petrol and diesel vehicles to CNG. We also have the world's only plant to convert natural gas to petrol. This plant, which was constructed to use up a surplus of natural gas, has since proved to be uneconomic, and is likely to be closed. ECNZ has converted three Daihatsu Mira cars to

Energy and Power

Energy: The standard SI unit of energy is the joule (J). A precise definition of this unit is given in the glossary.

In practical terms, the joule is approximately the energy it takes to lift a cup of coffee from a table to your mouth. Put another way, it would take about 100 kJ (100 000 J or 10^5 J) to heat that cup of coffee from cold to boiling in a microwave cooker. An active human needs about 10 000 000 J (10 MJ or 10^7 J) of energy for each day's total activity inclusive of breathing, heart beating, etc. There is 35 MJ of chemical potential energy stored in one litre of petrol, although usually only one quarter of this is extracted as useful energy by the car's engine. A car's fuel tank will store 1 GJ (10^9 J) when it contains 30 litres of fuel.

Power: Another unit which must be defined is the watt (W). This is a unit of power, the rate of changing energy from one form to another. In scientific terms, when energy is changed from one form into another, work is said to be done. For example, a 40 W incandescent light bulb changes 40 J of electrical energy into light and heat energy every second. Power, energy and time are inter-related by the equation:

$$\text{power} = \frac{\text{energy transferred}}{\text{time taken}}$$

We can do calculations with this information. For example, if the microwave cooker mentioned above is rated at 650 W, it supplies 650 J to our cold cup of coffee every second. If the coffee needs 100 000 J to be completely heated, we find that the

$$\text{time taken} = \frac{\text{energy transferred}}{\text{power}} = \frac{10^5}{650} = 154 \text{s}$$

or about two and one-half minutes.

Efficiency: In the case of the light bulb, less than 5 per cent of the electrical energy is converted into light. The rest of the energy supplied appears as heat and the light bulb gets hot! A fluorescent tube is considerably more efficient, converting about 25 per cent of the energy into light. In terms of energy used, fluorescent lighting illuminates a room to the same level as incandescent lighting for one fifth of the cost.

Efficiency is defined as

$$\frac{\text{useful energy}}{\text{total energy}}$$

and this fraction is often multiplied by 100 to give percentage efficiency.

$$\% \text{efficiency} = \frac{\text{useful energy}}{\text{total energy}} \times 100$$

A point to remember is that increasing the efficiency of our energy use is equivalent to electric cars, by replacing their petrol engines with electric motors and battery packs. As well as these major initiatives, individuals, aware of the problems looming, are working on alternative energy sources for personal transport.

Whatever the energy sources used in a country's development, each has economic and social costs, and environmental impacts. Therefore, the amount of energy that a country uses is a measure of its vulnerability to political and environmental changes. We feel that if New Zealand is to have a comfortable future, it will need to develop an energy mix which can be sustained, no matter what changes occur.

Over the last few decades, New Zealand's gross energy consumption (Fig. 1.3) has been rising, and it is reasonable to expect this rise to continue into the next century, as the population increases and industry expands, even if conservation measures reduce the per capita domestic energy consumption.

In this book, the word 'energy' is used in its widest sense. For example, we often refer to oil, wood and coal as 'energy', because these substances are stores of chemical potential energy which can be converted to useful mechanical or heat

---

obtaining extra energy supplies. Using the example of room illumination, converting from incandescent to fluorescent lighting releases electrical energy for other uses.

Because we are dealing with large amounts of energy, the joule is often not a conveniently sized unit to use. We can multiply the joule by powers of ten using prefixes, which produce the following larger units:

kilojoule	10^3 joules	(kJ)
megajoule	10^6 joules	(MJ)
gigajoule	10^9 joules	(GJ)
terajoule	10^{12} joules	(TJ)
petajoule	10^{15} joules	(PJ)
exajoule	10^{18} joules	(EJ)
zettajoule	10^{21} joules	(ZJ)
yottajoule	10^{24} joules	(YJ)

These prefixes can be applied to any unit. They are useful with power, and so in the text we use kilowatts (kW) and megawatts (MW) frequently.

Kilowatt-hours: Electrical energy is metered into the home and workplace in units of kilowatt-hours (kWh). Thus a one kilowatt heater would use 1 kWh every hour and a 100 W light bulb would use 100 W or 0.1 kWh every hour. At present, most domestic consumers pay close to 10c for each kWh of electrical energy supplied, which we will write as $10c(kWh)^{-1}$. This means that the heater costs 10c to run for one hour and the light bulb costs 1c. A kilowatt-hour implies energy use at the rate of 1000 joules every second ($1000\ Js^{-1}$) throughout an hour and this equates to 1000 x 3600 joules or 3.6 MJ.

1 unit = 1 kWh = 3.6 MJ which costs 10c (1994).

New Zealanders consume approximately 25×10^9 kWh of electrical energy every year (25 TWh y^{-1}), which would cost $2.5 billion if charged at the basic electrical energy rate. If you want to check your understanding at this point, you should be able to work out that the total electrical energy consumption is 90 PJ every year.

Because it is usual when discussing the generation, transmission and consumption of electrical energy to measure this in kilowatt-hours, we adopt this practice in the text, but display the equivalent quantity of energy in kilowatt-hours in brackets after the measurement in joules. For example, we use ECNZ's estimate for the annual energy requirement of a 'typical' home as 31 GJ (8 600 kWh). This is useful when we consider the interconversion of electrical energy with other energy forms, which are measured directly in joules.

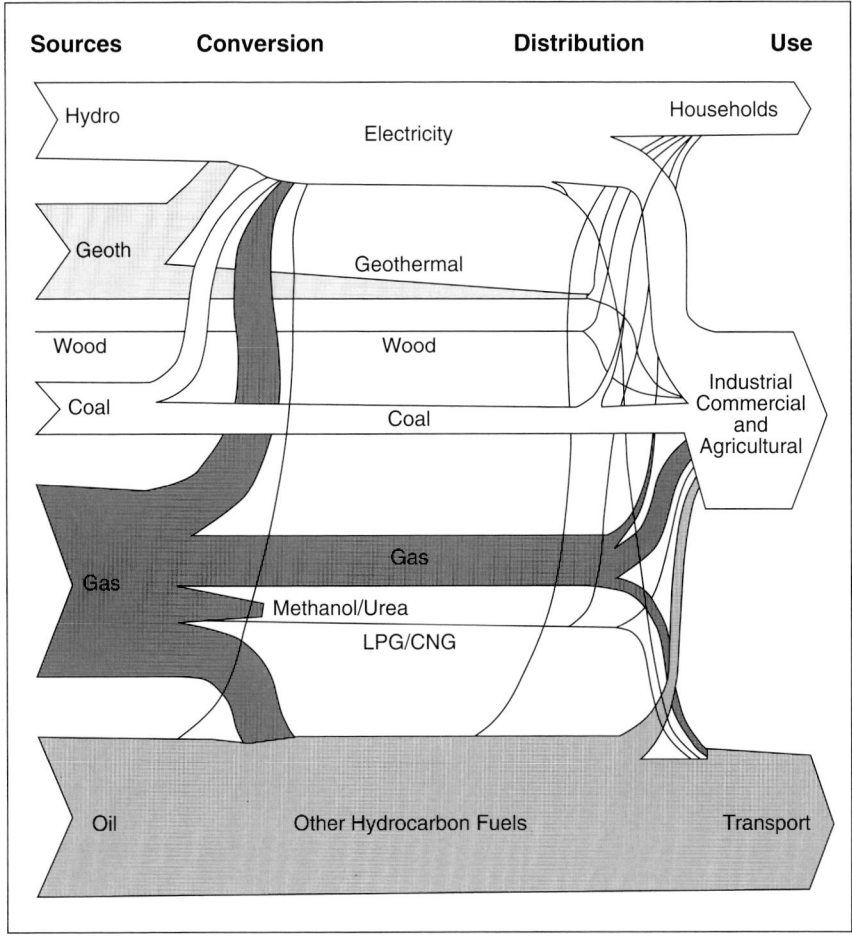

Figure 1.3
New Zealand's energy use in 1993. Primary sources (left) flow through conversion processes to consumers, with width of bands approximately to scale. (Ministry of Commerce)

energy by burning. The precise physical definitions of energy and power are given in the glossary.

The main forms in which energy is delivered to industrial and domestic consumers are: as combustible fuels, as electricity and as direct solar radiation. In this book we examine the supply, consumption and environmental impact of these forms of energy, and consider alternatives. We also discuss how New Zealand can adapt to the changing availability of energy supply, and how we can develop a sustainable energy policy, at the same time as we preserve our standard of living and our environment.

At present, New Zealand is not self-sufficient for its energy, being able to supply only 50 per cent of its liquid fuel and less than 80 per cent of its total energy from indigenous sources. The situation has deteriorated since 1990, when New Zealand was 81 per cent self-sufficient in terms of energy. Despite our having great reserves of coal and lignite, importation of crude oil is necessary to maintain our economy. A summary of the nation's energy supply and demand proportions is shown in Fig. 1.3.

This reliance on oil imports means that New Zealand is vulnerable to oil price rises or supply shortages caused by economic or political circumstances outside of our control. It is likely that within the next

Fossil Fuels and Carbon Dioxide Emissions

When carbon is burnt in a plentiful supply of oxygen the product is carbon dioxide.

$$C + O_2 = CO_2 \quad \Delta H^\circ = -394 \, kJmole^{-1}$$

This equation indicates that 12 g of carbon form 44 g of carbon dioxide, and that 394 kJ of energy are released during the reaction.

Scaling the reaction up, burning 12 tonnes (t) of carbon would liberate 44 t of carbon dioxide and make 394 GJ of energy available, or one tonne of carbon would produce 3.7 t of carbon dioxide and 33 GJ of energy. As both coal and charcoal are forms of carbon, we see that the combustion of both of these substances is a useful source of energy but that this combustion will release carbon dioxide to the atmosphere.

Climate scientists normally quote atmospheric carbon dioxide concentrations in terms of tonnes of carbon rather than carbon dioxide. Conversion may be made, as one tonne of atmospheric carbon is equivalent to 3.7 t of carbon dioxide.

Coal is a fossil fuel, having been formed over geological time by the action of heat and pressure on the remains of ancient plant material. Although the burning of coal contributes less than 10 per cent of energy production in New Zealand (Fig. 1.3), coal is used as a fuel for power stations in many countries. It is estimated that humankind has sufficient coal reserves for at least two centuries, although not all will be economically recoverable.

Natural gas or methane (CH_4) is another fossil fuel. This gas has been formed by the decay of organic material, and the planet's known reserves of the gas are estimated to be sufficient for 60 y. As the gas is produced by the anerobic decay of organic wastes, it is also formed in landfill sites and biogas plants as is described in Chapter 7. When methane, which is a hydrocarbon, burns in a plentiful supply of air, both carbon dioxide and water are formed.

$$CH_4 + 2O_2 = CO_2 + 2H_2O \quad \Delta H^\circ = -889 \, kJmole^{-1}$$

In this case, the combustion of 16 g of methane produces 44 g of carbon dioxide and 36 g of water, as well as 889 kJ of energy, much of this energy of combustion being derived from the formation of the carbon-hydrogen bonds of water. From this it can be seen that the use of methane as a fuel produces over twice the energy for the same emission of carbon dioxide. Converting a power station from coal to natural gas halves its carbon emissions.

Other hydrocarbons such as petrol also derive much of their energy of combustion from the formation of oxygen-hydrogen bonds and so produce fewer carbon emissions per kJ released. Assuming octane to be a typical component of petrol, the equation

$$C_8H_{18} + 12\tfrac{1}{2}O_2 = 8CO_2 + 9H_2O \quad \Delta H^\circ = -5464 \, kJmole^{-1}$$

shows that 114 g of octane produces 352 g of carbon dioxide on combustion as well as 5464 kJ of energy. This indicates that 14.25 g of octane produces 44 g of carbon dioxide, or contributes 12 g of carbon to the atmosphere, while making 685 kJ of energy available.

Liquified petroleum gas (lpg) is another hydrocarbon fuel. Its main component is butane (C_4H_{10}), which is kept liquid at room temperature under pressure in containers. When burnt it too forms carbon dioxide and water, as well as liberating energy, in this case 719 kJ for every 44 g of carbon dioxide produced.

$$C_4H_{10} + 6\tfrac{1}{2}O_2 = 4CO_2 + 5H_2O \quad \Delta H^\circ = -719 \, kJmole^{-1}$$

The formation of water as a by-product during the combustion of hydrocarbon fuels explains the large quantities of water which often issue from the exhaust pipes of cars when they are started on cold mornings, and the condensation which forms on the windows of houses which are heated by free standing lpg heaters. Examination of the above equations for the combustion of methane, butane and octane shows that in all cases the mass of water formed is greater than the mass of hydrocarbon burnt. As a rule of thumb, approximately one litre of water is formed for every litre of (liquid) hydrocarbon burnt.

The Transmission of Electrical Energy in New Zealand

On 1 July 1994, a state-owned enterprise, Trans Power New Zealand Limited, was created by the separation from ECNZ of the transmission facility. This SOE is charged with the transmission of electrical energy in New Zealand and distribution to electrical marketing utilities.

The national grid consists of a little over 12 000 km of high voltage AC, together with 600 km of high voltage DC link between Benmore in the South Island and Haywards near Wellington. This link includes the Cook Strait cable.

The grid has an 'average' carrying capacity of 5300 MW. Of this grid, a little over 8000 km is operated at 220 kV. It is important to maintain as high a voltage as practical in electrical transmission, to reduce the losses of energy which occur when a conductor is heated by the current flowing in it.

The electrical power P(W) or rate at which electrical energy is converted to heat energy in a cable with a potential difference V (V) between its ends and a current I (A) flowing within it, is

$$P = VI \qquad (1)$$

Now by Ohm's law

$$V = IR \qquad (2)$$

So we may substitute (2) into (1) to remove V (the potential difference across the ends of the wire, not the overall p.d. between the wire and ground) to give

$$P = I^2 R \qquad (3)$$

This equation tells us that in order to keep heating losses to a minimum, we should keep R and / or I^2 to as low a value as possible. To reduce R (Ω) we use aluminium cables of as large a diameter as is practical, given that the cables have to be hung from pylons. These cables have typical resistances in the order of 0.04 to 0.07 Ωkm[-1]. The best effect is obtained by keeping the current as low as possible, for the power losses vary as the square of the current, so halving the current reduces heating losses to one quarter of what they were.

Despite the low resistance per kilometre of the cable, the power losses are still significant. If we consider a cable carrying 22 MW, which is the power requirement of a medium-sized town, at a potential of 220 kV, then, by Ohm's law (2), the current will be 100 A. Using this current in equation (3) and a value of 0.04 Wkm[-1], we find that the power losses are 100[2] x 0.04 = 400 W for every kilometre of cable. In 1994 Trans Power reported that the loss ratio, the energy lost

Cook Strait Cable (see caption opposite)

over energy carried, of the system was 6 per cent which is in the order of 6 PJ of energy, equivalent to the output of a major hydro dam.

The main purpose of the high voltage DC link is the transmission of energy northward, although in the winter of 1992 energy was sent to the South Island. The link consists of a -350 kV /+270 kV transmission circuit. Direct current is used, as alternating current would result in losses in the submarine cable, because of inductive losses to the seawater surrounding the cable.

The cable has a maximum power carrying capacity of 1240 MW, with three cables in service at the time of writing. However, one of these cables is nearing the end of its economic life. If the cable were to fail, the capacity would be reduced to 1086 MW.

In the year ending 30 June 1994, 24.8 PJ (= 6883 GWh) of electrical energy was transferred to the North Island, a little over one third of the North Island's electrical energy requirement.

If, in the future, superconducting cables became available, and if the Cook Strait cable were to be replaced with superconducting cables, a much higher proportion of New Zealand's electricity could be generated in the South Island and transmitted northward.

twenty years the planet's reserves of oil and gas will become sufficiently depleted that the cost of crude oil will rise considerably in real terms as it did during the 1970s.

During the 'Think Big' programme of the 1980s, a Maui gas-to-methanol-to-synthetic-petrol plant was constructed in Taranaki; this supplies about one-third of our petrol demand. However this plant, although working as expected, is not efficient in energy terms, as it converts only about half of the energy available in the Maui gas into liquid fuel. Further plants of this type are unlikely to be built, even if more reserves of gas are found. It is likely that the Maui field's present reserves of gas, which stood at 3340 PJ in 1993, will become depleted by 2017, which is eight years after the present contracts expire. New Zealand continues to explore for both oil and gas, but at present no further major, easily recovered, reserves have been discovered.

During 1993, electrical generation provided 16.5 per cent of our primary energy production (Fig. 1.4). New Zealand has an installed electrical generating capacity of 7700 MW rated power, and this exceeds by 40 per cent the maximum power of 5500 MW that has ever been generated. The actual electrical surplus is much less than 40 per cent, as plants regularly need to be shut down for maintenance, and the water reserves in the hydro schemes are insufficient for continuous operation. ECNZ has stated that our present generation system could supply about 137 PJ (38 x 10^9 kWh) in one year, well above the 110 PJ (31 x 10^9 kWh) generated in 1993.

At present our demand for electricity is growing at 2 per cent per annum. This growth in demand is far below the predictions of 7 per cent per annum which were made some years ago, but the growth in demand could increase again for a number of reasons, which will be discussed later.

New Zealand obtains over 70 per cent of its electrical energy from its hydro dams. This form of energy is renewable, as is electrical energy obtained from the geothermal plants which contribute 5 per cent of our requirement. The remaining 25 per cent is generated by plants which use fossil fuels (Fig. 1.5).

If the country sustains growth in gross domestic product (GDP) of 3 per cent p.a., it is expected that demand for energy will grow at 1.4 per cent p.a. to be 50 per cent higher than at present by 2020. Electricity

Opposite: A section of the Cook Strait cable, showing the hollow gas duct and copper conductor at its core. The other layers are of insulation and protective material. (Photo: The Evening Post, Wellington)

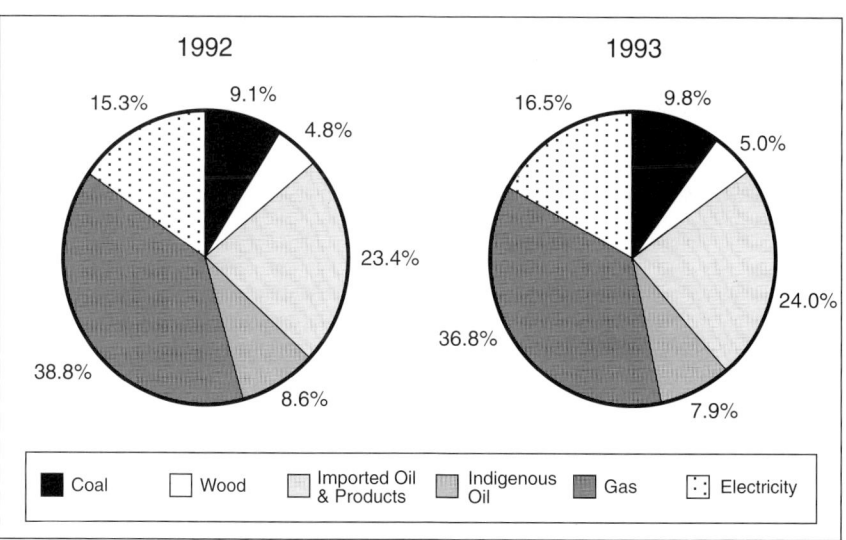

Figure 1.4 New Zealand's total primary energy production in 1992 and 1993. (Ministry of Commerce)

Figure 1.5
New Zealand's
electrical
energy sources.
(ECNZ Annual
Report)

Production		31.3.90	31.3.91	31.3.92	31.3.93	31.3.94*
Fossil fuels	GWh	6374	5643	6905	7774	9218
Geothermal	GWh	1805	2068	2065	2113	2681
Total thermal	GWh	8179	7711	8970	9887	11899
Hydro	GWh	20823	21845	21369	19682	27294
Wind turbine	GWh	–	–	–	–	1
Total generated	GWh	29002	29556	30339	29569	39194
Purchased	GWh	14	14	7	10	11
Total available	GWh	29016	29570	30346	29579	39205
Consumption						
Sales of electricity	GWh	27374	27892	28660	27753	36790
Transmission losses and internal use	GWh	1642	1678	1686	1826	2415
Total consumed	GWh	29016	29570	30346	29579	39205*

*This column refers to a period of 15 months to 31.3.94.

supplies will be under particular pressure, accelerating in turn the depletion of the Maui field.

Electricity costs seem certain to rise in real terms during this period, making several new methods of generation economic. If we use fossil fuels (oil, gas and coal) to meet the increase in demand, we must remember that these fuels, when burnt, produce gases which contribute oxides of carbon, sulphur and nitrogen to the atmosphere, increasing global warming trends, acid rain and photochemical smog.

It has been claimed that a coal-fired station such as that at Huntly releases more radioactive material (once trapped in coal) into the atmosphere than does a nuclear power plant during the same period of operation.

In a business-as-usual scenario, our emissions of carbon to the atmosphere would rise 62 per cent over a twenty-five year period, from our present 7 Mt to over 11 Mt, of which increased electrical energy production from fossil fuels would contribute 12 per cent.

Under international agreement (the Rio Convention on Climate Change), which New Zealand ratified in March 1994, we are committed to reducing the net release of carbon dioxide into the atmosphere to 1990 levels by the year 2000. Together with the international community, the country has to seek ways to decrease reliance on (non-replaceable) fossil fuels.

For the past thirty years, New Zealand's energy problems have stemmed from our needing to cope with a situation in which oil imports were becoming more costly and less reliable. Today, the energy problem that is most pressing is our need to reduce carbon dioxide emissions from the burning of all forms of fossil fuels.

Two main strategies suggest themselves for dealing with this latter problem. First, we need to increase the efficiency of use of fossil fuels. Second, we need to seek out alternative energy sources. These should not contribute greenhouse gases, and should be non-polluting, as well as economically and politically acceptable. Of course, they should also be energy sources that suit New Zealand's pattern of energy uses. A third strategy, of limited scale only, is the creation of a carbon sink by planting trees which are not allowed to

Superconductivity

Electrical conductors in general possess a property called resistance, which is the ratio of the potential difference (V) across the conductor to the current (I) flowing in that conductor.

This ratio $R = U/I$ has the units of ohms (Ω). For a large class of conductors, including metals, the resistance is constant for any value of current providing the temperature doesn't change. Although the resistance of the aluminium cables of the grid is quite low, being between 0.04 to 0.07, there is still a loss of electrical energy during transmission. The loss of electrical energy each second (P) may be calculated from the expression $P=I^2R$, and has units of power, which is the watt (W). The energy appears as heat, which means that the cable becomes warm during transmission.

In 1911 a Dutch physicist Onnes cooled mercury in liquid helium, a material he had first produced in 1908. He found that the resistance of the mercury became unmeasurably small below a critical temperature, and he named this phenomenon superconductivity. It is now known that the resistance of a superconductor is truly zero.

For many years, superconductivity remained of little commercial importance, as for materials to exhibit superconductivity they had to be cooled in liquid helium, which is an expensive material. Until 1986, the material with the highest known critical temperature was an alloy of niobium and germanium, Nb_3 Ge, which has a critical temperature of -250° C (23.2 K).

However, in that year researchers in Zurich discovered that a ceramic oxide of lanthanum, barium and copper became superconducting below -243° C (30 K). Since then the record critical temperature has been raised to over -148° C (125 K), which is well above the boiling point of liquid nitrogen. This means that it may be possible to transmit or store electrical energy by means of superconductors cooled in liquid nitrogen, which has a boiling point of -196° C (77 K) and is a cheap, safe and readily available material.

The transmission of energy would be revolutionised if a wire of superconducting material could be made. A superconducting cable could carry many times as much electricity as a copper wire of the same size. Underground transmission wires are normally immersed in oil, to cool the wire and maintain insulation. Replacing the oil with liquid nitrogen would not present any difficulties. The problem is to produce a flexible wire which could carry a high current density of, at least, 20 000 A cm^{-2}. The ceramics are brittle materials, unlike metals, and so are not easily woven into wire, but progress has been made and ribbons and wires up to 300m long have been produced (1994). These wires can conduct more that 10 000 A cm^{-2} when cooled in liquid nitrogen. In May 1995, workers at Los Alamos announced that they had succeeded in depositing the superconducting ceramic yttrium barium copper oxide onto zirconia coated nickel tape, to produce a superconductor capable of a current density of 1 million A cm^{-2}, which is over a thousand times that carried by ordinary copper wire.

About 6 per cent of electrical energy generated in New Zealand is lost in transmission. If the aluminium wires of the grid were replaced by superconducting wires, buried underground, and cooled by liquid nitrogen, the output of a major hydro dam would be saved. Even replacing the Benmore-Haywards link with superconducting cable could save 3 PJ, or about 3 per cent of yearly generation, and would make it possible to transmit more power northward, allowing a higher proportion of New Zealand's electricity to be generated in the South Island.

High-temperature superconductors would be useful for storing electrical energy in superconducting rings. Because a superconductor has zero resistance, once a current has been set up in a superconducting ring it flows without loss. In one experiment conducted in the United Kingdom in 1958, a current was maintained in a superconducting ring for 2.5 years. The current stopped only because a strike delayed the delivery of the liquid helium. It would be possible to bury giant superconducting rings near to cities, and store electrical energy in currents within the rings, extracting the energy when the demand was high. The engineering company Bechtel is building a prototype 20 MW storage ring, using conventional superconductors, near San Fransisco.

Superconductivity could revolutionise transportation. Not only are electric motors which use superconducting magnets highly efficient, but superconducting magnets could support levitating trains which run at speeds in excess of 300 kmh^{-1}, with no friction between track and train. Trains of this type are being developed in both Japan and Germany.

be burnt at a later date.

It is reasonable to predict that the next twenty years will see the introduction of private road vehicles which rely on stored electrical energy rather than liquid fuels, and that conditions will become favourable for the introduction of urban public transportation systems based on electrically propelled buses or trains. As transport constitutes 35 per cent of New Zealand's energy use (Fig. 3.1), this sector may place a heavy demand on our generating capacity. Therefore, in seeking to suggest pathways for New Zealand's energy future, systems which generate electrical energy may well be the most useful.

Another point to be made is that any new energy sources should be renewable, and that for political security they should be controlled within New Zealand. Fortunately, all of the sources which are available for adoption fit these two criteria. Data concerning these renewable energy options are summarised in Fig. 1.6.

Although energy prices in New Zealand have not increased to a point where they can be regarded as high by international standards (Fig. 1.7) it would be unwise for New Zealanders to ignore the potential changes that might occur within the next twenty years.

The purpose of this publication is to review these potential changes. Energy planning is a very long-term project, and market forces cannot be relied on to lead us to the best decisions from both the economic

Figure 1.6
New Zealand's
energy options:
current perform-
ance data.
(Ministry of
Commerce)

NZ$ (1992) IRR = 10%	Capital cost		Biofuel cost		O&M cost	Plant life	Plant factor	Overall cost
Energy source	$/KW	¢/kWh	$/GJ	¢/kWh	¢/kWh	Years	%	¢/kWh
Landfill gas [1]	2,000	4	-	-	1	15	85	5
Geothermal	4,000	5.5	-	-	0.5	30	85	6
Hydropower	3,500	8	-	-	0+	>50	50	8
Wind power	2,000	7.5	-	-	1.5	20	35	9
Wave-power	3,500	7	-	-	9	10	90	16
Solar heating - space	[2]	-	-	-	-	-	-	-
Solar water heating	[2]	13	-	-	1	20	30	14
Photovoltaic	12,000	59	-	-	1	30	30	60
Biofuel: [3]								
Woodchips & residues	2,400	4	2	3.5	1.5	15	80	9
Domestic firewood	200	11	2	1	-	10	3	11
Cereal straw	2,400	4	6	10.5	1.5	20	80	16
Biogas: [4]								
Processing wastes	1,000	1	-	-	1-	20	90	2-
Green crops	2,500	3	5	5+	3-	20	90	11

[1] Landfill gas collection costs included in capital and O&M. If excluded, costs fall by one-third.

[3] Cost of procuring biofuel varies with source: Arisings 2-5 $/GJ; Residues 0-2 $/GJ; Plantation fuelwood 2-4 $/GJ; Firewood 2-20 $/GJ.

[2] Inadequate information and methodology to develop comparative data.

[4] Digesting wastes to produce biogas saves power needed for aerobic treatment, and this is credited. Removing the credit increases costs by one-third.

Country	Regular petrol price c/litre	Rank	Domestic electricity price c/GJ	Rank
New Zealand	90	4	31	1=
United States	53	1	42	4
United Kingdom	135	5	55	5
Australia	82	3	38	3
Canada	72	2	31	1=
Germany	140	6	83	6
Japan	200	7	108	7

Figure 1.7
Some representative energy prices. (Ministry of Commerce)

and environmental points of view.

In particular, this book attempts to look beyond present-day expedience to a New Zealand which not only has sustainable growth into the twenty-first century, but which also meets its obligations to the international community, particularly to several of its near neighbours who would be adversely affected if the predicted effects of global warming were to come about.

If we have a future as a developed country, then we shall have to find a sustainable energy policy that works for us. So, too, if humankind is to continue to develop a global technologically advanced society for all, then a sustainable energy policy for the whole planet will have to be a part of that development.

2 The Need For Renewable Energy

Until recently, the main reason for the development of renewable energy has been the concern that the Earth's reserves of fossil fuels will run out in the near future.

At the time of the first oil shock in 1973, when crude oil prices rose by a factor of four, Sheik Youmeni argued against further rises, suggesting that the developed countries would rapidly find alternative energy sources, and cease to rely on importing oil from the Organisation of Petroleum Exporting Countries (OPEC).

During the last thirty years major reserves of oil and natural gas have been found, and our known reserves of fossil fuels have increased. At the present rate of use, the world has sufficient reserves of oil for at least forty years, as well as enough natural gas for sixty years, and coal for almost two centuries (Fig. 2.1).

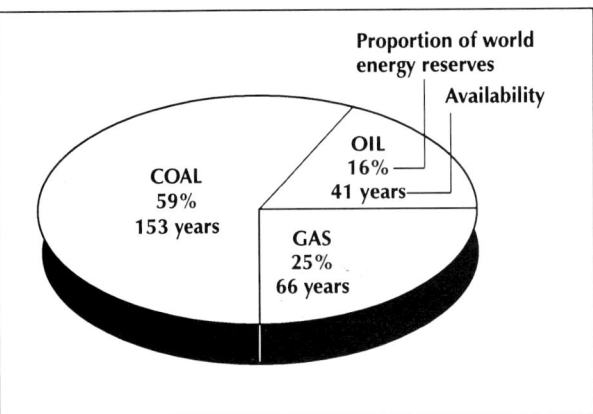

Figure 2.1
World primary fossil fuel reserves which can be extracted definitively and economically. (Shell, 1992)

It may seem that the pressure to develop and introduce renewable energy is no longer immediate, but another concern has recently become important. This is the possibility of global warming, caused by human enhancement of the greenhouse effect.

If there were no atmosphere, the Earth would be like the Moon, on which the surface temperature can be as high as 130° C at midday, and as low as -170° C during the night. A part of the reason why this doesn't happen here is the way in which heat is spread around the Earth by winds and ocean currents, particularly from the equator towards the poles. The combined effect is to moderate the overall surface temperature of the Earth. Even so, Vostok in Antarctica has recorded temperatures of -80° C.

It has been calculated that if the spreading of heat by winds and ocean currents were the only means of controlling the Earth's temperature, the average surface temperature of the Earth would be -6° C, which would make life difficult! At present it is 14° C. In the past, the Earth's average surface temperature has differed from this value. It may have been as high as 25° C 100 million years ago, during the Cretaceous Period, but as low as 10° C during the Ice Ages.

That the Earth's temperature is higher than we would expect if we considered just the effects of the winds and ocean currents, is due to the presence of 'greenhouse gases' such as carbon dioxide, methane, nitrous oxide (nitrogen(1) oxide), water vapour and some CFCs in the atmosphere.

Neglecting water vapour, our atmosphere consists of 78 per cent nitrogen, 21 per cent oxygen and 0.93 per cent argon. The final 0.07 per cent is composed of a variety of trace gases including these greenhouse gases. The atmosphere is transparent to visible light and to ultra-violet and infra-red light with wavelengths close to visible light. Much of the ultra-violet light is stopped by the ozone layer, which is rapidly becoming depleted due to the presence of man-made chlorine compounds which we have released into the atmosphere.

Of the radiation which strikes the Earth, about a fifth is reflected directly back into space, a little is absorbed by plants for use in photosynthesis, and the rest warms the oceans, the land and the atmosphere.

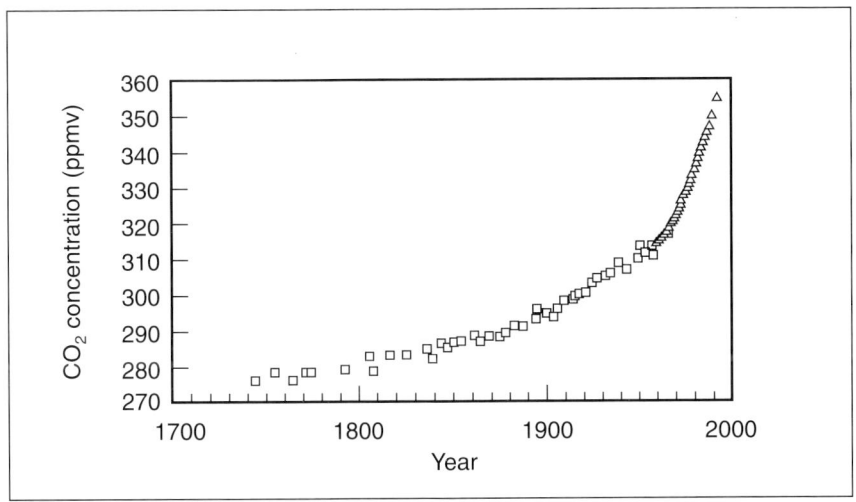

Figure 2.2
The increase of atmospheric carbon dioxide since 1700, showing measurements from ice cores in Antarctica (squares), and, since 1957, direct measurements from the Mauna Loa observatory in Hawaii (triangles). (After Houghton)

All objects emit radiation. The hotter the object is, the shorter the wavelength of the radiation. Our Sun, with a surface temperature of 5500°C (5800 K), emits mostly short-wavelength light, whereas the Earth, with an average temperature of 14°C (287 K), radiates long-wavelength infra-red light.

The Earth sends back to space as much energy as it receives, otherwise it would continuously get warmer. It is said to be in thermal equilibrium. As we have indicated earlier, if the Earth had no greenhouse gases in its atmosphere, thermal equilibrium would be reached at a temperature of -6°C (267 K).

The greenhouse gases absorb most of the longer wavelength infra-red radiation, so preventing this energy from being re-radiated into space, and maintaining our planet at 14°C rather than the chilly -6°C. In this way they act as a blanket, and keep our planet at a comfortable temperature for life. This greenhouse warming occurs independently of human activity and is the natural greenhouse effect.

In 1958 Charles Keeling started to measure carbon dioxide concentrations. At the time, the carbon dioxide concentration of the atmosphere was 320 parts per million (ppm). He noticed that the concentration was increasing each year. It is now over 350 ppm, which means that the atmosphere contains 2600 billion tonnes of carbon dioxide. As it is usual to quote this in terms of the amount of carbon rather than carbon dioxide, the figure is normally stated to be 700 billion tonnes (7×10^{11} t or 700 Gt) of carbon. Measurement of the carbon dioxide trapped in air-pockets inside the Antarctic ice sheet showed that before 1850 the concentration was 275 ppm (Fig. 2.2).

The increase in carbon dioxide concentration is due to human activity. Eighty per cent of the increase comes from the burning of fossil fuels, thereby releasing carbon which has been trapped underground for millions of years. The other 20 per cent comes from the destruction of rain forest, which is being cut down at the rate of 'a rugby paddock every three seconds'. The human race is at present releasing over 7 billion tonnes (7×10^9 tonne or 7 Gt) of carbon to the atmosphere every year, of which New Zealand contributes 0.1 per cent or 7 million tonnes, which is about 2 tonnes per head of population. In this respect, New Zealanders are more environmentally friendly than are the citizens of most developed nations. The United States emits over 5 t of carbon per head, with Canada and Australia not far behind. (Fig. 2.3).

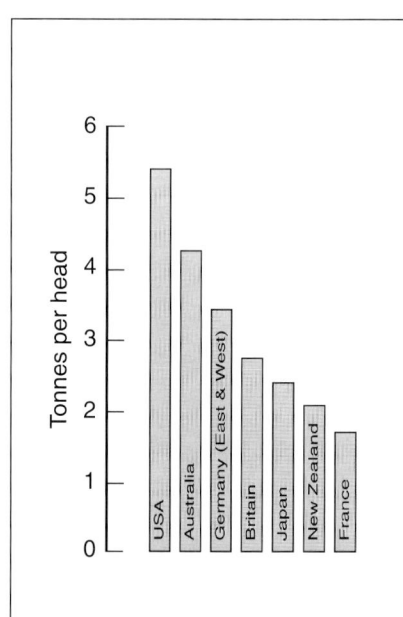

Figure 2.3
Emissions of
carbon dioxide in
1990.

Figure 2.4
Expected
atmospheric
carbon dioxide
concentration to
the year 2100.
(Greenpeace)

increase is keeping pace with the growth of human population. This is not surprising, as the main sources of atmospheric methane, other than its formation in wetlands and by nesting termites, are the coal, natural gas and oil industries, as well as agriculture and waste disposal. A small increase (from 280 to 310 ppb) has occurred in the concentration of nitrous oxide, mainly since 1950. In the case of the man-made CFCs, these did not exist prior to 1930. CFCs now occur in the atmosphere in amounts measured in parts per trillion (ppt); however, due to the damaging effect of certain CFCs on the ozone layer, their manufacture is being eliminated by international agreement (the Montreal Protocol). The damage caused to the ozone layer is quite unrelated to the greenhouse effect, and the two should not be confused. Discussion of ozone depletion is outside the scope of this book.

About half of the carbon released by human activity is removed because it dissolves in the oceans. Nevertheless, it is expected that carbon dioxide concentrations will lie between 450 and 550 ppm by 2050 and that, even if humanity restricts CO_2 emissions to 50 per cent of 1990 levels, the CO_2 level will still exceed 400 ppm by 2100 (Fig. 2.4).

Methane concentrations have also increased (Fig. 2.5). Pre-industrial levels, as measured from air trapped in Antarctic ice, were 700 parts per billion (ppb). These have now increased to 1800 ppb, and this

The greenhouse gases are not all equally effective at trapping infra-red radiation. The CFCs are very potent greenhouse gases, and even in concentrations measured in parts per trillion they have a measurable effect. Methane and nitrous oxide also make a significant contribution to the greenhouse effect in their present atmospheric concentrations. In fact, if we compare the gases in equal concentrations, carbon dioxide is the least effective of the greenhouse gases, methane being 7.5 times as potent, molecule for molecule. However, because its concentration is at least a hundred times as great as the other gases, carbon dioxide contributes 70 per cent of greenhouse warming, methane and nitrous oxide producing 23 per cent and 7 per cent of the warming respectively. It will be important to control all of the greenhouse gases, but at present the main effort to avert global warming is directed towards the control of carbon dioxide levels.

As the concentrations of greenhouse gases increase, we would expect their effect to increase, leading to a warmer planet. Climate records confirm that the planet has become warmer by 0.6°C since 1900 (Fig. 2.6).

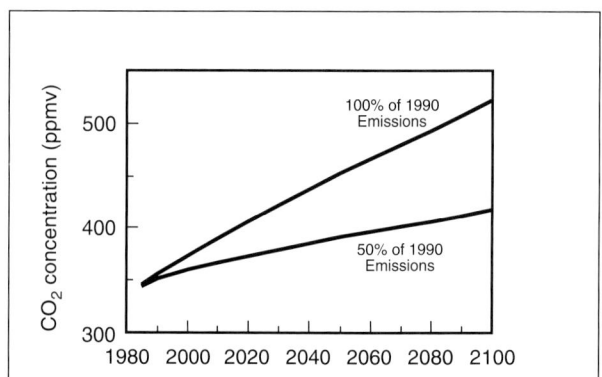

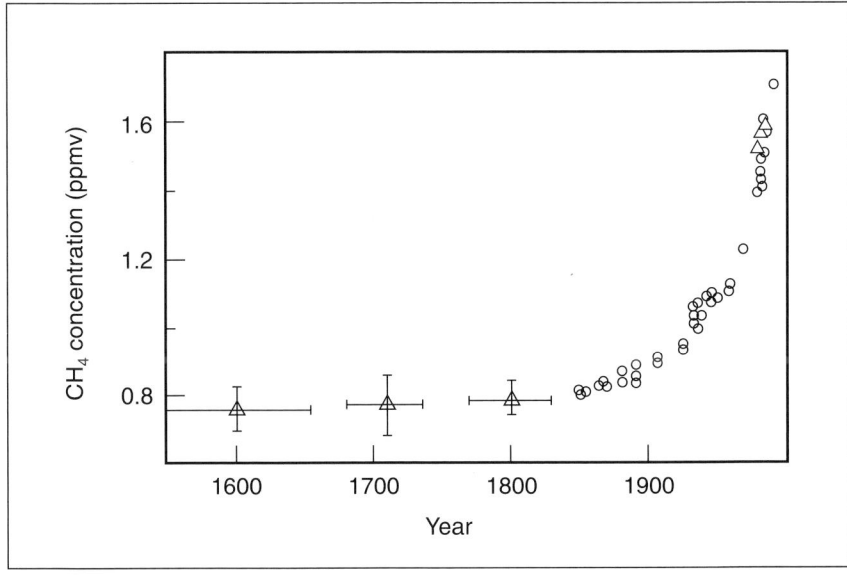

Figure 2.5
Concentration of
methane in the
atmosphere, 1600
to the present.
(After Houghton)

Whether or not this warming is a direct consequence of the increasing levels of greenhouse gases is still being researched, but most climate scientists now feel that it is. Certainly the eight hottest years (worldwide) on record have been recorded since 1980. The hottest year yet was 1990. Most temperature records have been kept for 140 years. It should be noted that the effect is one of generally warmer nights rather than days, which is consistent with the 'blanketing' effect of the greenhouse gases. An increase in the Earth's surface temperature caused by humankind adding greenhouse gases to the atmosphere is called the enhanced greenhouse effect.

If the warming is due to greenhouse effects and we continue on a 'business as usual' basis, it has been calculated that the Earth's climate will warm further, by between 0.5 and 2.0° C by 2030, and this warming will continue throughout the next century (Fig. 2.7).

At first sight this would appear to be a pleasant prospect. For example, the climate of Auckland would become like that of present-day Brisbane, and even Dunedin would enjoy temperatures more like those of Blenheim. So why are we concerned?

As the planet warms up, so will the surface waters of the oceans. This trend is already being observed. As the water warms, it expands, causing sea levels to rise. This is also happening and sea levels have risen 12 cm since 1900, although some of this increase has been due to melting of glaciers, and the removal of underground water for irrigation. The sea levels

Figure 2.6
Global and New
Zealand temperatures since the
1850s. (After
Salinger)

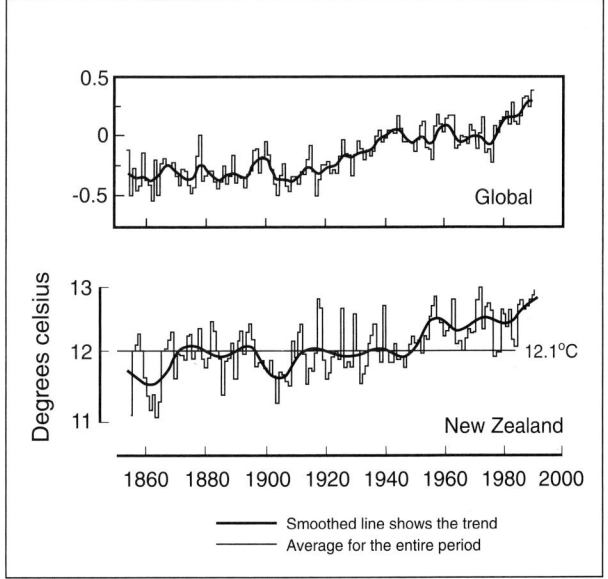

are expected to rise another 25 cm by 2030.

It is also possible that the polar ice-caps will begin to melt, and if this happens the sea levels will rise many metres. Indeed the complete melting of the ice-caps would cause sea levels to rise 60 m. It is not known what is happening to the ice-caps at present. Athough the Greenland ice-cap seems to be getting smaller, the Antarctic ice-cap may even be getting larger, because as the planet warms, more moisture will evaporate from the ocean surface, and this may lead to greater precipitation over Antarctica, depositing more ice there. It

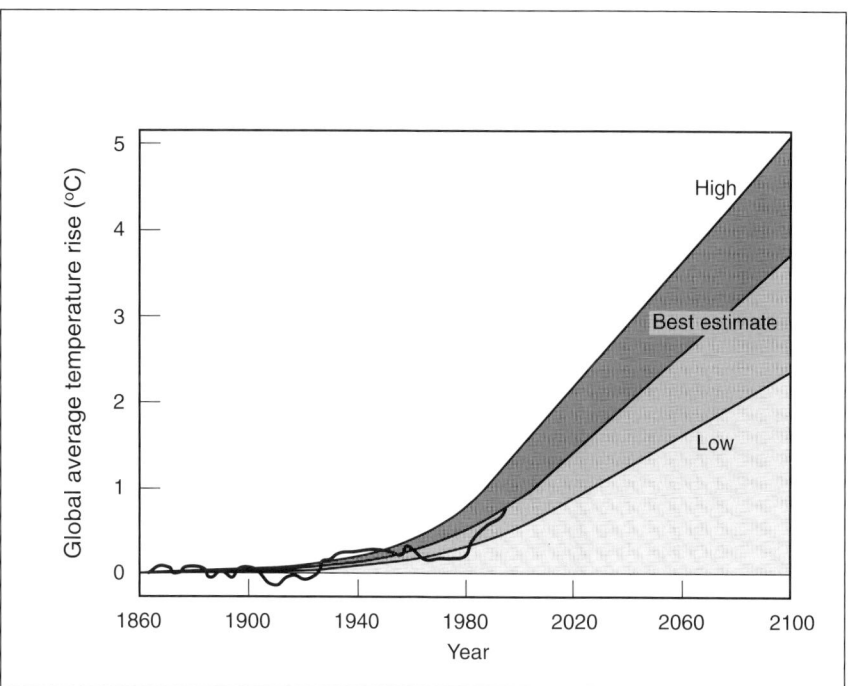

Figure 2.7
The Earth's
climate will warm
during the next
century.

would certainly require at least one hundred years for substantial melting of the ice-caps to occur.

Many of the Earth's cities and areas of valuable farmland lie at sea level. There is a danger of catastrophic floods if the sea levels rise significantly. It is even possible that Pacific Island countries such as Tonga could eventually become totally submerged. Whatever happens, it is certain that if sea levels continue to rise, many countries, including New Zealand, will have to spend large sums of money protecting themselves against rising water.

Another predicted effect is that climatic belts will shift as the planet warms, and agricultural production will have to change. This will benefit some people. For example, wheat production in Russia will be favoured. It is expected that North America and parts of Australia, which together produce much of the world's wheat, will get less rainfall and will not be suitable for wheat production. It is difficult to predict the effect, taking the whole planet into account, but if agricultural production were disrupted, then famines might well follow, and this might lead to increased political instability and even wars, which could affect New Zealand, or in any case would be most undesirable. A change in climatic regions could also mean that diseases such as malaria, which at present are not established within New Zealand, may well become a problem to us.

New Zealand's economy, and the affluence of its citizens, relies to a great extent on agricultural production. Again, it is difficult to predict the effect that changing climate will have on us, but as an example it could be expected that kiwifruit and apple production would have to move southwards, which might cause hardship for orchardists established in the northern regions, and that tropical pests and diseases might become established.

Probably of most serious immediate concern is the likelihood that weather patterns will become more violent, with stronger winds, more intense rains and longer periods of drought world-wide. Effects of this type are already being noticed. The Netherlands have suffered major flooding each winter during the years 1992-5, and in 1993 the United States experienced not only 'the Storm of the Century' during March, but also major flooding in the mid-west. The average wave height in the North Atlantic has increased by 50 per cent compared with the 1960s, with storm waves being 10 per cent higher. Further evidence comes from the insurance industry. From 1987 to 1994 there have been fifteen events which have cost the insurance companies more than one billion (US) dollars (in terms of constant 1990 dollars), after a period of twenty years without any billion-dollar claims. The US$15 billion claim caused by Hurricane Andrew in 1992 holds the record for any single event. This claim may have mounted to US$75 billion if the hurricane had struck Miami, thirty kilometres to the north. The German re-insurance organisation Munich Re has supplied a graph illustrating this trend (Fig. 2.8).

Although this evidence has to be treated cautiously, considering the increasing skills of claims lawyers and the higher affluence and population densities of some of the areas affected by storms, it is being presented at international conferences without serious challenge. The trillion-dollar insurance industry is certainly taking the evidence seriously, for although it has reserves measured in the tens of billions of dollars, if a major claim were to remove these reserves the insurance industry could be brought close to bankruptcy, which would disrupt world trade, and affect every human.

As the greenhouse effect has such serious implications for the human race, governments have begun to take action. The Enquete Commission, which advises the German Government, reported in 1991 that 'our planet is already warming at an increasing rate'. Other governments have received similar information and in 1992 the Framework Convention on Climate Change was negotiated. This has now been ratified by the required number of countries. It states that governments must have adopted, by 21 September 1994, policies to stabilise carbon dioxide emissions at their 1990 levels, by the year 2000.

New Zealand is a party to this convention, and it is our task to meet the requirements of the Framework Convention. To do this, we have to find ways to reduce our present dependence on fossil fuels. As one of the developed nations, we are more responsible for the emissions of greenhouse gases than those countries that are still developing. It has been said that Australia has a greater greenhouse effect than does India, and New Zealand's contribution would not be far behind. Put bluntly, if we wish to maintain, or increase, our standard of living, then we will have to adopt a sustainable energy policy, and if we wish to avoid the unpleasant effects of global warming, then we cannot leave it to the other countries.

Figure 2.8
Great natural disasters, 1960-1992: trend of insured losses. (Munich Re)

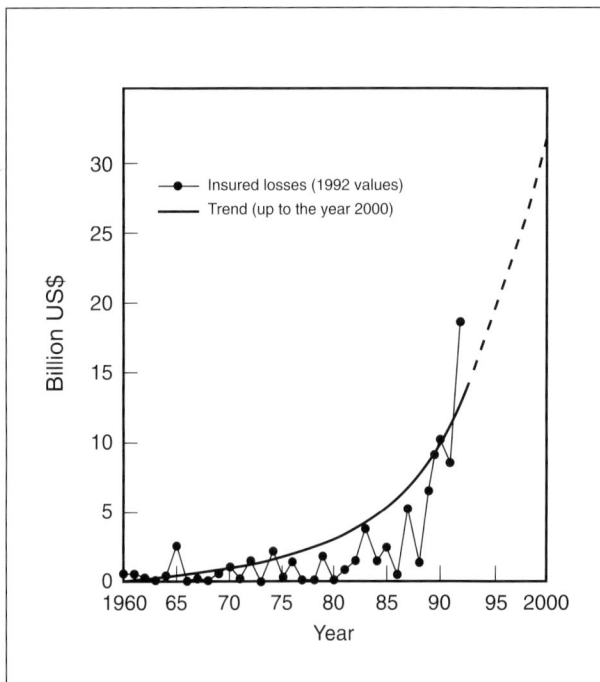

3 Energy Efficiency and Conservation

In 1993, the Ministry of Commerce and ECNZ released new projections for energy supply and demand. Although these covered a range of different growths in GDP, all indicated a steady growth in energy demand, and in particular, a need for commencing the construction of more electricity generating plant before the turn of the century.

An example was a 'business-as-usual' projection based on a GDP growth of 3 per cent over the next twenty-five years. This indicated that energy demand would grow at 1.4 per cent annually, becoming 50 per cent higher than today's demand by 2020. In all cases, the cost of energy was forecast to rise. This would reduce our international competitiveness, unless industry could operate with greater energy efficiency.

If the increase in energy demand were to be met by the use of fossil fuels, this would have profound implications for New Zealand. It is possible that the demand could be met by accelerating the use of Maui gas, which at present supplies 35 per cent of total demand. This would occur if the proposed 400 MW combined cycle-gas turbine power station were to be built at Stratford. Alternatively we could make use of our coal deposits, which could meet demand into the twenty-second century. The Ministry of Commerce estimated, in its 'business-as-usual' scenario, that with an annual GDP growth of 3 per cent, our CO_2 emissions would rise by 62 per cent over twenty-five years, with 12 per cent of this increase coming from increased electricity production.

As we have shown in the section on global warming in Chapter 2, such an increase in CO_2 production is unacceptable. The UN Conference on Environment and Development (the Rio Earth Summit) in June 1992 proposed, in the Framework Convention on Climate Change, that countries stabilise their CO_2 emissions at 1990 levels by the year 2000. New Zealand ratified this in March 1994 and so, together with the rest of the world, we must make efforts to meet this target. It is expected that we will do so by reducing our CO_2 emissions to achieve 20 per cent of the requirement, at the same time increasing the CO_2 uptake in our forests to meet the other 80 per cent. A rational strategy for our country would see a move from fossil fuels towards renewable energies, as we increase the efficiency of our energy consumption.

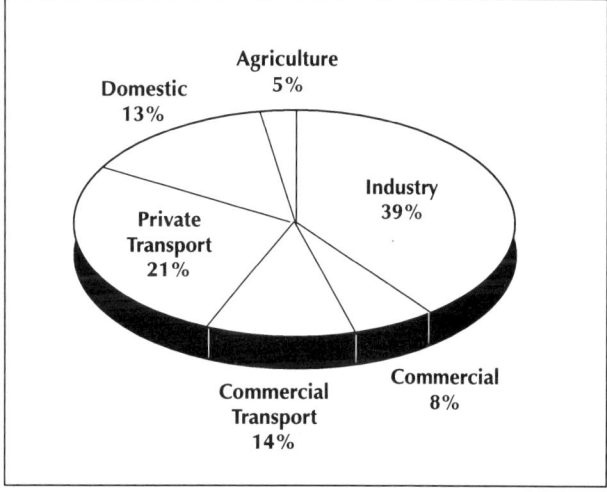

In New Zealand, energy use can be classified according to the sector of society in which it is used. The pie chart (Fig. 3.1) indicates the percentage use in each sector. The transport sector, which at present requires 35 per cent of our energy supply, is expected to experience a growth of 1.8 per cent each year. By comparison, the industrial and commercial sector is expected to grow at 1 per cent per annum, and the domestic sector at 0.7 per cent. As the transport sector represents the major part of the projected growth, potential savings in this area are discussed in a separate chapter (Chapter 12).

Energy efficiency means we obtain the

Figure 3.1 Energy use in New Zealand by sector (year ended June 1993). (Energy Efficiency and Conservation Authority)

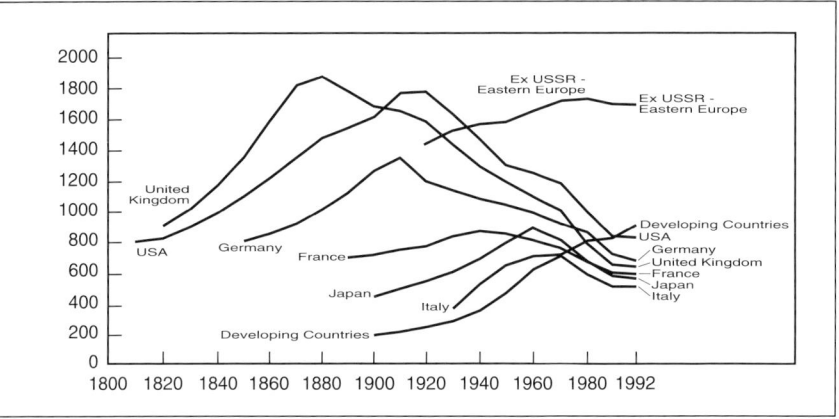

*Figure 3.2
The amount of
energy required
to produce each
unit of GNP is
declining in
developed
countries and
increasing in
developing
countries. (World
Renewable
Energy Confer-
ence 1994)*

The Progressive Pricing of Electricity

The pricing of electricity in New Zealand is complex, and both politically and commercially controversial. While it is beyond the scope of this book to discuss all the details of the price structure, here is a brief overview:

Utilities purchase their power from ECNZ essentially on two scales. The baseline scale for guaranteed consumption is at an agreed-to price for the maximum load each quarter of a year. Additional electricity is then purchased as required on the spot market at prices which vary from as low as 2 cents per kWh during summer nights to about 18 cents per kWh for those winter periods when the hydro storage lakes are low.

Unlike in many industrialised countries, the domestic electricity consumption in NZ is proportionally very high at about 40% of the total consumption, while 50% is consumed by industrial and commercial users and 10% is lost in transmission. Electricity is the preferred form of energy for domestic water and space heating, and for cooking. While the domestic price of natural gas in the North Island is about 4 cents per kWh, compared to 10 cents per kWh for electricity, the South Island domestic gas price is about 15 cents per kWh compared to 9 cents per kWh for electricity. Monthly line charges for electric grid connection and delivery charges for piped or bottled gas are additional. Reduced electricity rates for ripple-controlled hot water heaters, night storage heaters and general domestic night-time domestic consumption are available from most distribution utilities. Standard commercial rates are a little higher than domestic rates, but special arrangements are made with very large consumers. Some special consumers such as churches and sports clubs, and also domestic consumers who are a long distance from strong grid connections, pay much higher than average rates.

Interested readers might like to approach their local supply authority to determine the method of pricing to consumers operating in their region.

One factor which can encourage conservation of energy is a pricing system which rewards the careful use of energy. At present, nearly all New Zealand consumers of electrical energy pay a supply charge for connection to the electrical system. This charge can be a significant proportion of the cost for many households. The electrical consumption is then charged at a flat rate of about 10 c(kWh)$^{-1}$ Many feel that this system does not encourage the conservation of electrical energy.

If conservation of electrical energy is considered important, progressive pricing would be helpful. Progressive pricing means that consumers pay a low charge for their first block of electricity, more for the next block, and so on.

New Zealand has a considerable resource of hydro dams which have been in operation services we require, such as lighting, warmth, manufacturing and transport, with the use of less energy. At present the costs of energy supplies are increasing, while the cost of energy efficiency technology is decreasing. This means that efficiency must lead to financial gain, both in the home and in industry, bringing greater profits and a higher standard of living. Most of the industrialised countries have made significant improvements in energy efficiency, while maintaining economic growth. Between 1973 and 1985,

Japan's GNP grew by 46 per cent while its energy consumption hardly changed, and in Denmark the amount of energy required to produce each unit of GNP, which is termed the energy intensity, has fallen to 70 per cent of its 1972 value. Fig. 3.2 illustrates this trend for several countries.

Geoff Bertram of Victoria University has shown (Fig. 3.3) that with respect to energy intensity the New Zealand economy has lagged behind other OECD countries since 1982, and at present is performing more like the economy of a developing country.

In a speech in 1994, the Minister of Energy the Hon. Doug Kidd said that New Zealand industry could make energy savings of up to 20 per cent, and that this would represent gains of $500 million each year (one third of the cost of the Clyde Dam!). As industry uses 39 per cent of our energy, this would represent an 8 per cent saving nationally, postponing for several years the need for new generating plant. To give specific examples, the Lion Breweries bottling plant in Newmarket has achieved savings of $150,000 each year since 1992, mainly by careful control of the refrigeration plant and lighting; Wilson and Horton are saving $90,000 each year, having modernised their factory air-con-

for many years, and which produce electricity at a cost which is possibly as low as 1.17 c(kWh)$^{-1}$. If new plant were to be commissioned by ECNZ or by a private group, the energy might cost as much as 11c(kWh)$^{-1}$. At present ECNZ averages the costs of generation and sells electrical energy wholesale at an average of about 5c(kWh)$^{-1}$, the costs of transmission and distribution bringing this to the price charged to the consumer. Because ECNZ can average prices, it is difficult for competitors to construct new generating capacity, as they would have to sell electricity at the real cost. Progressive pricing would reflect the true cost of generating each block of electricity, and make it more possible for generating capacity to be constructed by groups other than ECNZ.

If New Zealand's pricing policy were to be based on the concept of a low supply charge, a low initial charge for a basic number of kWh, which could be a significant fraction of the 8600 kWh purchased by the 'average' New Zealand household, and a progressively increasing charge for each block of energy over this, then energy consumption might be reduced. Not only would consumers be reluctant to 'waste energy' if this energy were paid for at the top end of the range, but also the 'payback times' of energy conservation measures, such as house insulation and the installation of heat pumps, would be reduced.

*Figure 3.3
Energy intensity indexes for the New Zealand economy compared with OECD average.
(Geoff Bertram)*

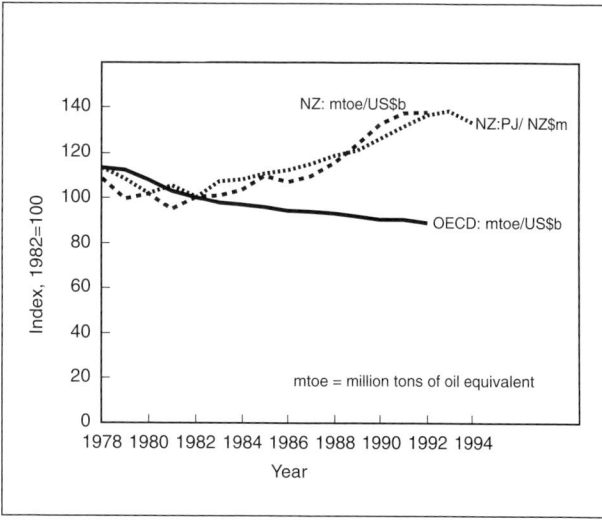

ditioning; and at 70 The Terrace, Wellington, the running costs of the building have
been cut by $48,000 a year, by fitting high
efficiency lighting.

Energy efficiency strategies are also
profitable for the energy suppliers. If the
supply companies sell their customers the
means to increase energy efficiency, and
assist in the installation, then these companies can obtain extra profits. The company still makes a profit on the sale of the
electricity, without needing to invest in
new generating capacity, and also profits
from the sale of the equipment.

As well as industry, the public sector can
introduce efficiencies. There are over 4000
institutions, such as hospitals and schools,
in which the energy use is split between
heating and lighting, with a small use by
machinery. Considerable savings can be
made in these institutions merely by turn-

ing down, or off, the heating and lighting
in rooms which are not occupied. The
National Library Building in Wellington
used over $1.1 million of electricity in
1987, its first year of use. In 1994, the cost
of electricity for this building had been
reduced to $550,000. This was achieved
by care taken in the nature of lighting employed, the installation of timer switches,
and some modifications to the heating and
ventilation systems.

Householders and their families may
also benefit from practising energy conservation. Although household fuel and
power may cost a little less than $1000 out
of an 'average' family budget of almost
$32,000 (Fig. 3.4), lower income house-
holders spend similar sums, and so a
higher proportion of their income goes on
heating and lighting.

As there are 1 200 000 households in

Figure 3.4
Energy expendi-
ture in an average
family budget of
$31,720 (New
Zealand Yearbook
1992).

Food	Housing	Fuel & power	Other operations	Clothing	Transport	Other goods	Other services
17%	21%	3%	10%	5%	18%	11%	14%
$5512	$6604	$884	$3380	$1456	$5824	$3536	$4524

New Zealand, even a $50 saving each year in each household would represent a national saving of $60 million, or in energy terms, over 2 PJ, which is about one-quarter the yearly output of a major hydro dam.

Looking at the breakdown of electrical energy use in a New Zealand household (Fig. 3.5), we can see that on average nearly half of the household's energy consumption is used to heat water, and this represents a yearly cost of almost $400, or $450 million nationally. It is in this area that significant savings may be made.

Possibly the greatest energy inefficiency in the home is heat energy lost from water heaters, either because of old, poorly insulated hot water cylinders, or because the cylinder's thermostat is set at too high a temperature. It has been estimated that if all domestic hot water cylinders were lagged and the thermostat set at no more than 55°C, the electricity used for domestic hot water would be reduced by 13 per cent and the national electricity demand by 2 per cent. These steps alone would save the householder $60 each year, meaning that the payback time for cylinder wraps for lagging is about one year.

Greater savings could be made if house-holds used either heat pumps or solar water heaters for domestic hot water. Solar water heaters, such as the Sola60 or the Thermocell unit, cost about $3000 to install, although it must be remembered that if fitted to a new house the cost of a standard domestic hot water cylinder, which is $600, can be set against this cost. Tests in the Auckland area show that savings of over 10 GJ (3 000 kWh) per household are possible. This is an annual saving of $300, and represents a tax-free return of 12.5 per cent on the investment, with a payback time of less than ten years. As the life of a cylinder is at least twenty years, these cylinders make economic sense even in less favoured areas of the country.

Heat pumps are also able to effect considerable savings. Although they cost $4000 to $6000 to install, they can still produce savings of about $300 per annum, representing a 6 per cent tax-free return.

Space heating	Water heating	Lighting	Cooking	Refrigeration	Other
24%	42%	9%	8%	10%	7%

Figure 3.5 Electrical energy use in a New Zealand household. (Data: Ministry of Commerce)

House in Tauranga, with a 3m² 'Sola60' thermo-syphon solar water heater on its roof. (Photo: Sola60 NZ Ltd)

Thermocell solar water heating panels built into the roof of a Christchurch home. Total area is 4.8 m². When combined with a wet-back log burner, the system can virtually eliminate water heating electricity costs. (Photo: Thermocell Ltd)

The heat pump can also be used for space heating, increasing the yearly energy savings of the household.

It has been estimated that solar water panels could save 8 per cent of New Zealand's electricity consumption, and that heat pumps could save 5 per cent, amounting to a total saving of 13 per cent. Heat pumps improve the effective use of energy in electrical space-heating by a factor of about three. However, it must be noted that as the efficiency of fossil-fuelled power stations is in the order of one third, the use of a heat pump only brings the situation back to where it would be if the fossil fuel were to be burned directly in the home. If care is taken to eliminate atmospheric pollution, domestic solid fuel heaters represent an efficient way to heat homes. It is more effective to heat a house by the direct burning of solid fuels, or gas, in a space heater, than it is to burn the fuel in a power station at 33 per cent efficiency, transmit the electrical energy with some loss, and then use it in resistive heating.

Ninety-seven per cent of New Zealand households have washing machines, which can be major users of energy. Nearly all of the energy used in a hot wash goes into heating the water. Nowadays, there are very effective cold water detergents: about 60 per cent of washing powder sales are of these. Savings of up to 1.0 per cent of domestic electricity can be achieved by the elimination of hot-water washing.

The Heat Pump

A heat pump can be thought of as a refrigerator working backwards. A refrigerator removes heat energy from the inside of a box and disposes it outside, using some energy, usually electrical, to run a compressor pump. In the case of a heat pump, heat energy is removed from outside and brought into the box, which is the building, or whatever is being heated. Again, some energy will be used in the process, and this energy will also appear as heat, adding to the heating effect. The overall effect of a heat pump is that the outside temperature is reduced, while the inside temperature is increased.

We will consider how a heat pump may be used to heat a room. A heat pump consists of a compressor unit which compresses a working fluid, often a CFC. When something is compressed, work is done compressing that material and the work appears as heat. A fine example of this effect is seen when a bicycle pump is used. As the air is compressed by the pump, the end of the pump becomes quite warm, due to the work being done by whoever is using the pump. After compression, the fluid which has now become hot is allowed to lose heat to the interior of the room which we wish to heat.

Commercial application of Quantum hot water cylinders. The evaporator panels on the roof draw heat in from the atmosphere. (Photo: Tasman Energy Quantum Ltd)

The compressor motor will also convert a little electrical energy to heat, and this heat may be released into the room as well.

When the working fluid has lost its excess heat to the room, it is allowed to flow to an expansion chamber outside of the room. Here it is allowed to expand through a fine jet and it will change to a gas. When a substance changes from a liquid to a gas, the molecules of the substance become much further apart, and energy has to be supplied to overcome the attraction that the molecules have for one another. This energy can only come from the thermal (heat) energy of the gas and so the gas cools down. The energy involved is known as latent heat of evaporation.

In this way the working fluid beyond the expansion chamber has become quite cold. It is allowed to warm up outside of the room, extracting heat from the outside air, which is, of course, cooled down in the process. The warmed working fluid then flows back to the compressor, and the process is repeated.

There is a problem here, in that if the temperature of the outside air is close to, or below, 0° C, then the warming process does not occur efficiently, and the performance of the heat pump falls off considerably. Yet it is when the temperature of the outside air is low that heating is most wanted.

The advantage of the heat pump over ordinary (resistive) electrical heating is that with resistive heating, although the electrical energy is converted to heat energy with 100 per cent efficiency, nevertheless a megajoule of electrical energy cannot produce more than a megajoule of heat. In the case of the heat pump, a megajoule of electrical energy might bring a further two megajoules of heat energy into the system, giving three megajoules of heat energy. In this way, the heat pump has three times the heating effect of resistive heating. As an example of how this improvement can be used, a 650-watt 'Quantum' water heater manufac-tured by Tasman Energy Ltd of Richmond (Nelson) can heat water more quickly than can a 2-kilowatt standard water heater. Heat pumps are manufactured in New Zealand by both Tasman Energy and Carrier Airconditioning (NZ) Ltd.

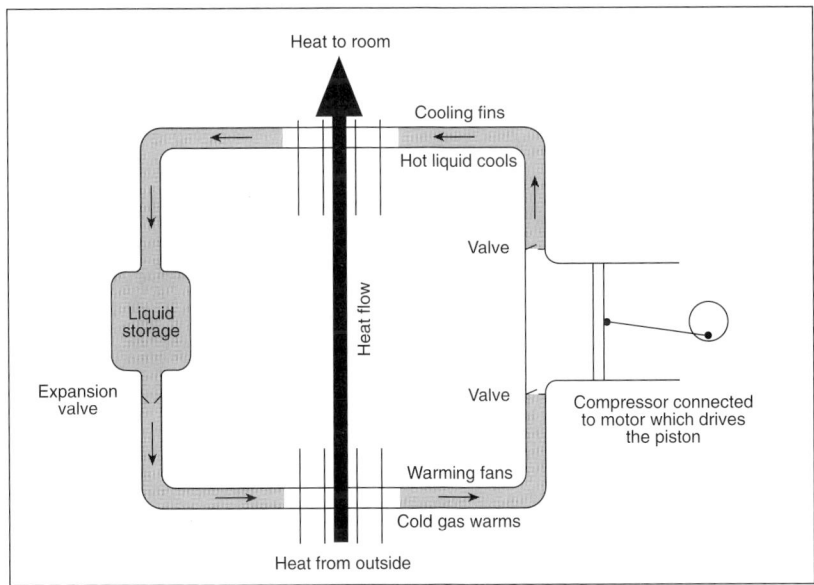

Figure 3.6 The Basic Design of a Heat Pump.

Domestic space heating contributes one-quarter of the typical household's energy bill. Here the most cost-effective saving is the fitting of insulation. Covering the ceiling area with a layer of fibreglass or wool insulation can reduce the heat loss by a third, and save the cost of the insulation in under ten years. Even the fitting of draught-strips and good curtains into older dwellings can make these perceptibly more comfortable in cold windy weather, and it is now possible to fit coated window glass, which reduces the outwards transmission of infra-red radiation through windows, an avenue of major heat loss in homes. The US Department of Energy has shown that 'Energy Advantage' glass contributes up to 2.5 times more heat to a home than it loses during a 24-hour period averaged over the year. They claim that each average-sized window using this glass could save about 35 GJ or 10 000 kWh in a twenty-year period. This would reduce energy bills by NZ$50 per window each

Carbon Taxes

One of the problems, comparing the costs of the various energy forms available to us, is that the calculated costs rarely take into account the environmental costs of generating the energy. This became apparent during the 1980s, when both the US and the UK generated large quantities of electrical energy from coal-fired power stations. The coal that was used in these power stations contained sulphur compounds, and so when it was burnt large quantities of sulphur dioxide were given off into the atmosphere. The power stations were equipped with high smoke stacks, which dispersed the sulphur dioxide and carbon dioxide produced, and prevented pollution of the surrounding area. Nevertheless the sulphur dioxide was eventually converted into sulphuric acid by oxidation,

$$SO_2 + H_2O + (O) \Rightarrow H_2SO_4$$

and this sulphuric acid was carried by prevailing westerly winds towards countries to the east, where it fell as acid rain, causing damage to forests, lakes, agriculture and buildings. In particular, Canada, Eastern Europe and Scandinavia have suffered, both economically and environmentally, from the acid rain. It took intense political pressure from the affected countries to cause the offending countries to modify their power stations to prevent the release of sulphur dioxide, and even today the ecosystems of many countries have still not recovered.

Although the effects of acid rain are now well understood, and international agreements exist to control the release of oxides of sulphur, many countries are now concerned that the increase in atmospheric carbon dioxide concentration might lead, through global warming, to increasing storm damage and higher sea levels. As we have said in the main text, this might cause severe economic and social damage to countries which contribute little or nothing to the carbon dioxide levels. In particular, several Pacific Island countries might become uninhabitable, possibly even disappearing beneath the ocean, as water levels rise. As the increase in atmospheric carbon dioxide levels is a global problem, the question is who will pay for this damage?

Recently, the concept of a carbon tax has been proposed. This is a tax which charges the carbon dioxide producer for the estimated cost of that pollution to society. The money raised could then be used to offset the social and economic costs of the pollution, or to fund international research into renewable energies. If the fund were to be managed by an international body, it would lead to the possibility of the whole world co-operating to solve a global problem.

This tax would raise the cost of electricity generated from fossil fuels and encourage the use of non-polluting systems such as the renewable resources.

It is difficult to estimate the social cost of a pollutant with any certainty. However, a study undertaken in California made the proposals summarised in Fig. 3.7:

year – a good return on the cost of installation.

For those of us who live in the southern
half of the South Island, it is a little surprising to see the low percentage of electricity consumption given in Fig. 3.5 for
household space heating. Although no
detailed, region-by-region, statistical information appears to be available at present,
it is believed that in Otago and Southland
space heating is likely to represent more
of the domestic electrical load than does

water heating, especially as 'nite-store'
heaters and other night-rate electrical heating systems are particularly popular in
these regions.

The use of such devices is essentially
market-driven. The present pricing policy
of electricity (see 'The Progressive Pricing of Electricity', p. 28) involves some
cheap 'spot-price' electrical energy being
available to utilities, especially at night.
Thus by offering essentially half-price
units of electrical energy between 11 pm

Nitrogen Oxides	Sulphur Oxides	PM 10*	ROG**	CO_2
$11600 /tonne	$11500 /tonne	$7800 /tonne	$3300 /tonne	$7 /tonne

Fig. 3.7 Estimates of the social costs of
various pollutants ($US).

* Particulate matter less than 10 microns
** Reactive organic gases

The effect of such a carbon tax would be to
increase the price of coal, liquid fuels derived
from crude oil, CNG, LPG, and electrical
energy generated from fossil fuels. Because
the various fossil fuels make differing
contributions to CO_2 emissions (Fig. 3.8), the
carbon tax levied on each fossil fuel would
differ. Coal would carry the heaviest tax. The
renewable energy forms would benefit, and
the cost of the use of these energy forms
would reflect better their environmental
advantages with respect to fossil fuels.

Coal 1.2	Oil 0.92	Natural gas 0.71

Fig. 3.8 Contributions of fossil fuels to CO_2
emission (kg CO_2 $(kWh)^{-1}$).

Research on the effects of moving from our
current taxation structure to one based on a
carbon tax was carried out in New Zealand by
Berl and Simon Terry Associates. Their report,
published in 1993, contained the following
conclusion:

'Simply collecting existing tax revenues on
the basis of carbon content (carbon tax) could
be expected to reduce CO_2 emissions by 15-
20% and lower energy intensity by 10-12 %.'
In other words, if we were to abandon income
tax, GST, and other forms of taxation in favour

of carbon taxes, we could lower our CO_2
emissions considerably. A second
conclusion contained in this report stated:
'A CO_2 tax of $213.59 $tonne^{-1}$ of carbon
was sufficient to allow a fall in GST rate
from 12.5% to 10%. There was only a
slight fall in environmental gains, and the
model reported a gain in GDP of 1.4%
and 33,000 extra jobs while projecting a
26% reduction in carbon emissions and
an 18% fall in energy intensity, relative to
"business as usual".'

Although carbon taxes seem likely to
benefit the New Zealand economy,
studies in Europe have highlighted the
potential drawbacks of such a tax. There
will be practical difficulties in ensuring
that polluters do not evade the tax, but
more importantly these taxes will hit the
poor more than the better-off. Because
the poor (and the elderly) spend a greater
proportion of their income on domestic
heating, and often live in less well-
insulated houses, an increase in the price
of either electricity or of fossil fuels for
home heating will disadvantage this
group in society. Because many of New
Zealand's homes are heated by electricity
generated by hydro-schemes this is not so
likely to be a problem here.

and 7 am, distribution utilities have encouraged domestic consumers, and some commercial customers, to use resistive wire electrical background heating of buildings in a big way. In the southern half of the South Island, such devices tend to be left switched on for over six months of the year. If this cheap electricity came entirely from hydro-generation, using water which would otherwise be released over dam spillways, then obviously it would be a good way of utilising energy which would otherwise go to waste. If, however, it involves the overnight running of fossil-fuelled generating equipment, which would otherwise be shut down, then it is not a good option.

New building regulations recognise the need for energy effectiveness in dwellings. These regulations are not always met. It has been calculated that if the code were carefully enforced, 0.15 per cent of the total New Zealand electricity use could be saved over the next twenty years.

Other domestic savings could be made by the replacement of old refrigerators and freezers by modern units (which may be as much as 40 per cent more efficient), by the greater use of microwave cookers, and by the replacement of incandescent lighting by fluorescent lighting, especially now that the compact, bayonet-fitting types are available.

As any energy saved in the home represents more disposable income for the occupants, it is clear that it is in each person's interest, as well as the interest of the nation and of the planet, to take the simple steps necessary to reduce domestic energy use.

4 The Storage of Electrical Energy

One of the great difficulties in the supply of electrical energy is the lack of a simple means of large-scale storage, whether as electricity or in a readily convertible energy form. If electrical energy is going to be used for personal transport or to be generated domestically by photovoltaic cells mounted on the roofs of houses, then some relatively cheap, safe and portable means of long-term storage must be found.

The most successful way of directly storing electricity is as a current loop in a superconductor. However, this requires very high technology, due to the nature of the materials and the low temperatures involved. For commercial purposes, this form of storage is beyond our present technology, although a prototype system is under construction near San Francisco.

It is easiest to store electrical energy as some other form of energy, which may be converted to electrical energy on demand. The potential energy of water in a hydro dam represents one such store of energy which may readily be converted to electrical energy.

Until now the most common method of storing electrical energy has been the rechargeable electrochemical cell. As cells are often joined together into a 'battery' of cells, the device is known as a battery.

The standard battery used in vehicles and many other applications is the rechargeable lead-acid 'accumulator' which was discovered in 1859 and which has been continually developed over the twentieth century. For automotive use, the battery usually has six cells each delivering 2 V. This gives a nominal voltage of 12 V. Each cell consists of a negative electrode of lead, a positive electrode containing lead dioxide, and an electrolyte of sulphuric acid.

The lead-acid battery has many disadvantages. Lead components are toxic, and sulphuric acid is a very corrosive liquid. This means that if the battery spills, considerable material and environmental damage can occur. The disposal of used lead-acid batteries and the production of lead and its compounds presents environmental problems. Standard car batteries do not maintain their energy storage capacity if they are often fully discharged or left for long periods in a discharged state. This disadvantage has been partially overcome in the modern deep cycle battery. The lead-acid battery does not operate well in low temperatures, and this has proved to be a problem with General Motors' prototype electric car, the Impact, which is now only being tested in the southern United States. However, the major drawback of the lead-acid battery is its low energy storage density.

It is possible that the lead-acid battery can be improved. Electrosource of Austin, Texas has developed a battery which

Figure 4.1
The lead-acid accumulator battery. (Courtesy, Rendle)

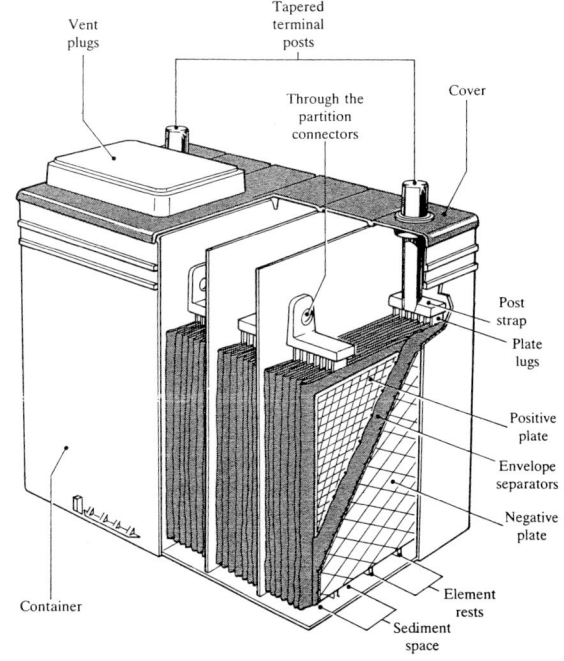

weighs 40 per cent less than conventional batteries. This battery, which uses lead-coated fibreglass electrodes, can withstand over 900 charge cycles; it is claimed that a battery pack able to power a minivan and costing about $5000 could be developed.

The standard car battery weighs about 18 kg and will probably have an energy storage capacity of forty amp-hours (Ah), which means that it can sustain a current of 1 A for forty hours. This converts into an energy storage capacity of 12 x 1 x 40 x 3600 J or 1.7 MJ, which is about 0.5 kWh. To store the energy of one litre of petrol using standard car batteries, would require about twenty batteries or 360 kg by weight.

The problem is not as bad as it might seem, for electric motors are much more efficient than the internal combustion engine and, particularly in the case of a commuter car, an electric motor does not use energy when the car is halted in traffic, although the heater, radio and lights would continue to drain the batteries. The electrically powered BMW E2 concept car has a design top speed in excess of 100 km/h and a maximum range of 420 km, with a practical range of 260 km. This car relies on an advanced sodium-nickel chloride battery pack of 200 kg. Similar prototype cars such as the General Motors Impact, which relies on a 500 kg lead-acid battery pack, are expected to be ready for production by 1998.

An alternative battery system, the Nicad battery, uses a cadmium positive electrode and a nickel oxide negative electrode in concentrated (alkaline) potassium hydroxide electrolyte. Although produced in a sealed, maintenance-free package, for the consumer electronic market, the high toxicity of cadmium and its components have raised environmental concerns. A 200 kg Nicad battery pack has been used in the Nissan FEV concept car, which has roof-mounted solar cells to augment charging. The Nicad batteries used in the FEV have a very low internal resistance, and are thin so that any heat produced during charging is quickly released. This allows the FEV's battery pack to be recharged in fifteen minutes. Both the FEV and the Impact use regenerative braking, whereby during braking some of the car's kinetic energy is returned to electrical energy stored in the battery pack.

If electric cars are to enter production, then better electrical storage systems are essential. It is clear that internal combustion engines which burn fossil fuels will have to be phased out in time, to prevent the increasing pollution of cities and further rises in the atmospheric concentration of carbon dioxide. In the first move towards this phasing-out, California has passed an act by which 2 per cent of the city's vehicles must become zero emission (i.e. electric) by 1998; this proportion must increase to 10 per cent by 2003. The need

The electrically powered BMW E2 concept car. (Photo: BMW AG)

exists for better energy storage systems, and several advanced rechargeable batteries are in the development stage. There is a requirement for batteries with long lifetimes, high energy storage density, charging efficiency and peak power output.

The nickel-metal hydride battery is one such promising system. The battery acts by reversibly absorbing and releasing hydrogen in the following reaction:

$$NiO(OH) + MH \Leftrightarrow Ni(OH)_2 + M$$

where M is a metal such as lanthanum. It can store 50 per cent more energy by weight compared to a lead-acid system, and is environmentally more sound. A battery of this type developed in the United States has withstood over 1000 charging cycles, which represents more than three years' use. It has produced the highest peak power-to-weight ratio for any battery system, equivalent to over 60 kW from a mass of batteries similar to that of a standard car engine. In 1994 the Geo Metro, a prototype passenger car powered by these batteries, travelled 345 km on a single charge.

Other batteries, such as the sodium-sulphur and lithium-polymer batteries, have the capacity to store up to five times the energy by weight of the lead-acid system. The sodium-sulphur battery is close to large-scale production for electric vehicle applications. The battery uses a negative electrode of sodium and a positive electrode of sulphur; its chemistry involves the reversible formation of sodium polysulphides:

$$2Na + xS \Leftrightarrow NaS_x$$

The battery operates at a temperature of 350° C, at which both electrode materials are liquid. The electrodes are kept apart by a ceramic composed of aluminium oxide, through which sodium ions may migrate. To obtain best performance, the battery has to be heated to 350° C prior to use. This is not difficult, for once brought to 350°C the internal electric current keeps the battery at its operating temperature (with some loss of useful energy). The main problem with the cell relates to safety issues, such as having a large mass of batteries, albeit well insulated, at 350°C and the very corrosive nature of the hot materials.

Nevertheless, sodium-sulphur modules are installed in the Ford Ecostar, of which eighty units have been prepared for public testing. This electric vehicle is based on the Escort van. It carries 350 kg of sodium-sulphur batteries, has a 450 kg payload, with a designed top speed of 120 kmh^{-1} and a range of 160 km at 70 kmh^{-1}. One of the prototypes came first in an endurance rally, the 'Tour de Sol, USA' during May 1994, covering 302 km on one charge; another has been used successfully by a British police force. Although two of the prototype vans have experienced fires, it is expected that the van will be commercially available in 1998.

BMW have modified eight BMW 325iX cars to electric power (Fig. 4.2), using sodium-sulphur batteries. These cars have been used in Munich and Nuremberg for delivering letters and parcels. They have cruising ranges in city traffic of 150 km, with top speeds of 100 kmh^{-1}. A comparison of the performance of their sodium-sulphur battery system with some of the alternatives is shown in Fig. 4.3.

The Japanese research organisation NEDO has developed a sodium-sulphur battery module for the storage of energy during off-peak hours. The module has a power output of 1000 kW over eight hours, which is an energy storage potential of 29 GJ, its overall efficiency in terms of AC input/output is 70 per cent and the module has an expected service life of 1500 cycles, which gives it a probable life of ten years. The module is being developed to allow for full utilisation of energy from nuclear power plants, but it would be ideal for the storage of energy in remote locations, where energy might be generated from either wind turbines or photovoltaic arrays.

The lithium-polymer battery is a solid-state system, which stores energy by the trapping or 'intercalation' of lithium ions in its electrodes. When an ion moves from one electrode where it is weakly bonded, to the other where it is more strongly bound, energy is liberated. The reverse process results in the storage of energy. The chemical nature of the electrodes and electrolyte does not change, rather the lithium ions move backwards and forwards in a 'rocking chair' fashion. Because the movement of the small lithium ions does not cause major changes in the electrodes, these do not disintegrate after repeated charge-discharge cycles, and it is expected that lithium-polymer batteries will have much longer lives than more conventional batteries. For the future, the lithium-polymer system represents a rugged battery tolerant to shock, vibration or deformation in an accident. It is leakproof, does not give off gases, and can be built into complex shapes to fit vehicle design. The

Sodium-sulphur battery pilot plant, developed by NEDO, the Japanese new energy research organisation. (Photo: NEDO, Japan)

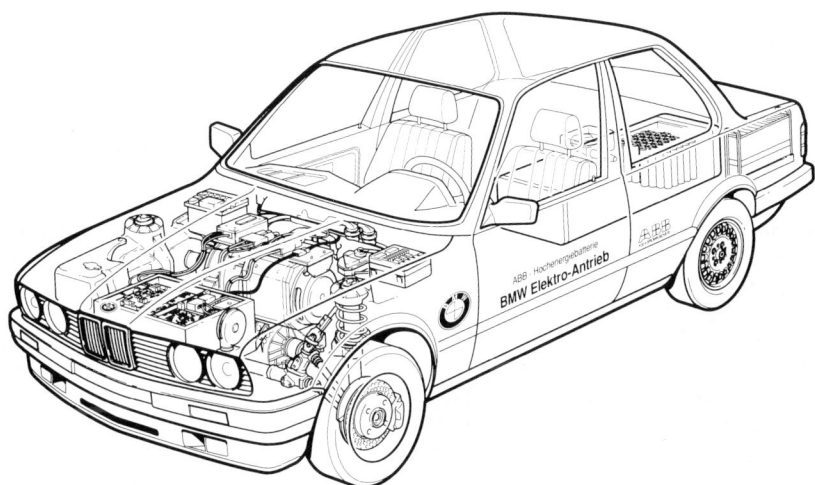

Figure 4.2
A BMW car
modified to
electric power,
using sodium-
sulphur batteries.

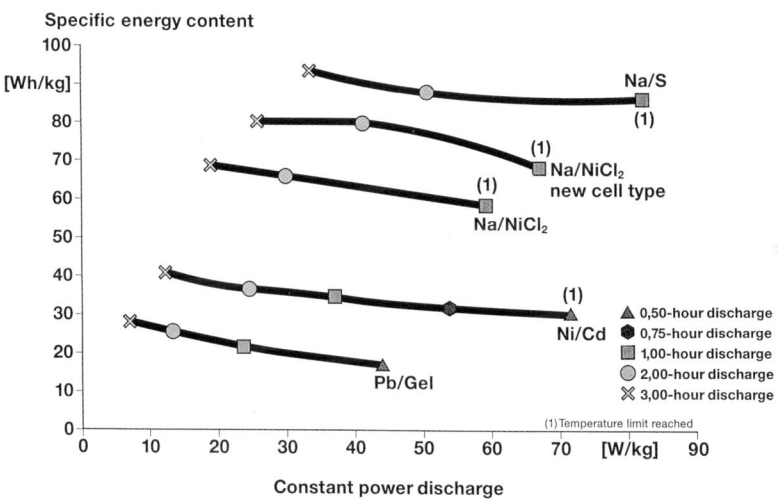

Figure 4.3
Comparison of
battery systems.

batteries are already in use in 'smart' credit cards, portable computers, and consumer electronic devices, but it is unlikely that large capacity lithium-polymer batteries can be available within ten years. In terms of energy storage, this battery has a similar energy storage density to the sodium sulphur cell, and can store four times as much energy per kilogram than the lead-acid battery.

Other battery systems are under development, but the sodium-sulphur, nickel-metal hydride and lithium-polymer batteries offer good prospects of making electrical energy storage practical, whether for electrically propelled personal transport or for domestic storage.

In the longer term, energy may be stored in the form of hydrogen. This gas, together with oxygen, is produced when water is electrolysed, and may be stored either in the form of compressed gas, liquid (boiling point -253° C or 20 K) or adsorbed in some metals such as nickel. When the

Hydrogen as a Fuel

Hydrogen represents a concentrated store of energy. When burnt, one kilogram of hydrogen liberates 143 MJ of energy, over four times that of a kilogram of petrol:

$$H_{2(g)} + \tfrac{1}{2}O_{2(g)} \Rightarrow H_2O_{(l)} \qquad \Delta H = -286 kJ\ mole^{-1}$$

Hydrogen is difficult to store. A tonne of hydrogen gas has a volume of 12 million litres, which is about the size of a large school assembly hall. Even when compressed into cylinders, hydrogen is not a useful fuel. A 40-tonne truck loaded with compressed hydrogen cylinders only transports half a tonne of hydrogen and hence much less energy than the same truck loaded with petrol.

Hydrogen can be condensed to a liquid at -253°C (20 K), and a tonne of the liquid has a volume of 14 thousand litres. A road tanker could transport several tonnes of liquid hydrogen. The Space Shuttle, which burns liquefied hydrogen and oxygen in its main engines, is among the few vehicles which currently use hydrogen. NASA has road and rail tankers which carry 5 tonnes of liquid hydrogen at a time, and stores 3 million litres of liquid hydrogen at Cape Canaveral.

Hydrogen may be generated by the electrolysis of water, when hydrogen is produced at the negative electrode:

$$H_{(aq)}^{+} + e^{-} \Rightarrow \tfrac{1}{2}H_{2(g)}$$

The process can be operated with an efficiency over 90 per cent, but as it uses electricity the hydrogen produced must inevitably be more expensive than the electricity used. Even so, if hydrogen were produced when there was excess electrical production, such as at night, in the case of nuclear reactors, or in the day in the case of PV cells, the energy is effectively stored rather than wasted. The hydrogen could then be used in a fuel cell when electricity was needed.

Hydrogen-fuelled cars developed by BMW. (Photo: BMW AG)

stored energy is required, the hydrogen may be used either in a fuel cell (see below) to produce electrical energy directly, or it may be used in an internal combustion engine. BMW are developing hydrogen-fuelled cars, and despite the difficulties of dealing with liquid hydrogen have produced a car which runs on hydrogen gas; it may be refuelled in a matter of minutes, unlike battery-powered cars, which usually require several hours for recharging. As the sole product of the combustion of hydrogen is water, the car is non-polluting. It has been suggested that liquefied hydrogen, which is used in the Space Shuttle's main engines, would be a suitable fuel for aviation purposes, and that during the next century, commercial aircraft will be propelled by this fuel.

The Hydrogen Fuel Cell

Hydrogen may be oxidised in such a manner that the energy of the reaction appears mostly in the form of electrical energy, rather than heat:

$$H_{2(g)} + \tfrac{1}{2}O_{2(g)} \Rightarrow H_2O_{(l)} \quad \Delta H = -286 \text{ kJ mole}^{-1}$$

The reaction shows that 8 g of hydrogen (4 moles of hydrogen gas) will liberate more than 1 MJ of energy, or that 250 g hydrogen will liberate as much energy (36 MJ) when oxidised as 1 litre (~ 750 g) of gasoline.

The hydrogen fuel cell consists of a vessel in which two platinum electrodes are immersed in a potassium hydroxide electrolyte. Hydrogen is fed to one electrode where oxidation occurs:

$$H_{2(g)} \Rightarrow 2H^+_{(aq)} + 2e^-$$

At the other electrode oxygen is reduced:

$$\tfrac{1}{2}O_{2(g)} + 2e^- \Rightarrow O^{2-}_{(aq)}$$

The product of the reaction is pure water, which means that the hydrogen fuel cell is a non-polluting energy source.

The electron flow is from the hydrogen electrode to the oxygen electrode, so in terms of conventional current flow the hydrogen electrode is the negative electrode.

Practical fuel cells were first developed for the US space programme, when a convenient supply of electrical energy was required for the Apollo missions, and it became clear that batteries were inadequate for the task. Fuels other than hydrogen can be used in fuel cells. The Solid Oxide Fuel Cell can use hydrogen, methane or methanol. Methane is the main component of Maui gas, and of landfill gas, and has already been used in a compressed form as the transport fuel CNG. Methanol is a liquid fuel which could be made in New Zealand from either Maui gas or wood residues. As a fuel cell-electric motor unit may achieve over 40 per cent efficiency, and it is expected that this could be increased to 50 per cent by the year 2000, fuel cells represent a more efficient use of fuels than the internal combustion engine.

Fuel	kJ per gram
Hydrogen gas (H_2)	143
Methane gas (CH_4)	56
Petrol (octane, C_8H_{18})	48
Coal (carbon, C)	33
Ethanol (C_2H_5OH)	30
Methanol (CH_3OH)	23
Carbohydrates (eg. $C_6H_{12}O_6$)	16
Carbon monoxide gas (CO)	10
Water (H_2O) and carbon dioxide (CO_2)	

The energy ladder: each figure shows the kilojoules that are produced when a gram of fuel is burnt.

5 Wind Energy

Over the past twenty years much progress has been made towards harnessing wind energy. A wide range of wind turbines are in commercial production and may be bought 'off the shelf'. These machines are reliable, efficient and effective devices for electrical generation and commercial grid-connected wind farms have been erected in several European countries, and in California. Denmark, which has led the world in this technology, now produces over 3 per cent of its electricity from wind energy, and hopes to be producing 10 per cent from wind farms by the year 2000. Germany has now exceeded Denmark in installed capacity. A number of Asian countries also have sizeable development programmes, and Australia opened its first wind farm in 1994.

Wind is caused by the heating of our planet by the Sun. Between latitudes 30° and 60° there exists a belt of strong westerly winds (Fig. 5.1). For New Zealand, which lies between 34° and 47° S and is one of the few land masses in the circumpolar Southern Ocean, these are the winds that the nineteenth century sailors christened 'The Roaring Forties'.

Wind turbines operate effectively over a wide range of wind speeds. As Fig. 5.2 shows, a typical turbine such as the Vestas V27, which has been operating in Wellington since March 1993, develops its rated power, which in the case of the V27 is 225 kW, at a wind speed of 12 ms⁻¹ (43 kmh⁻¹). The turbine, which is not designed to produce more than 225 kW, continues to develop this rated power as the wind speed increases. The power output is regulated by altering the pitch of the blades. To avoid

Figure 5.1 Global circulation of wind over the earth.

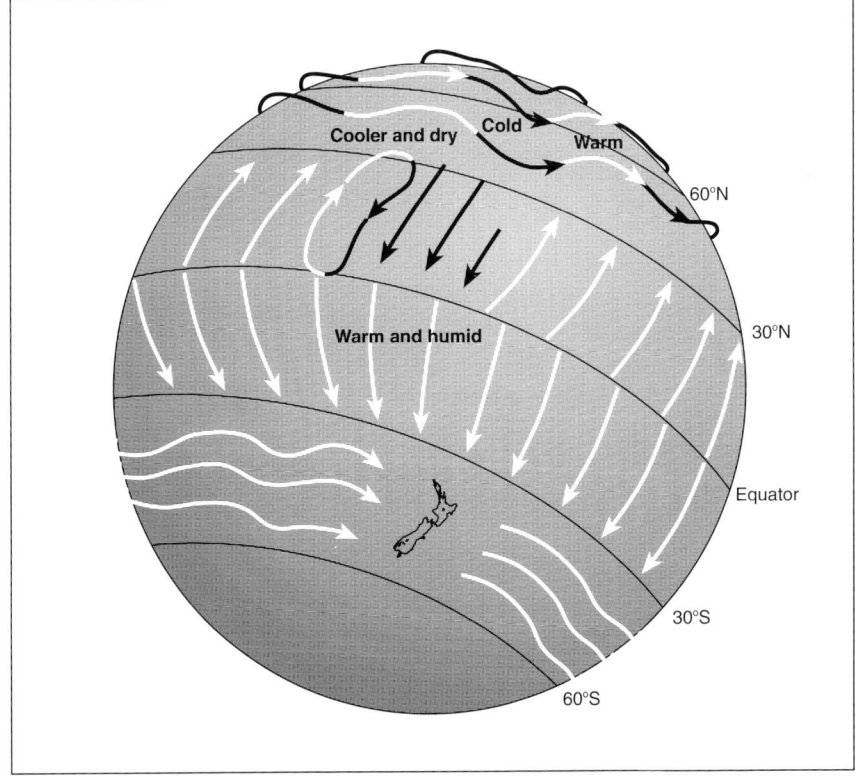

Capacity Factor

When discussing the economics of renewable energies, it is important to consider the Capacity Factor. This is simply the

$$\frac{\text{Energy produced in a certain time}}{\text{Energy that would be produced if the plant ran at full capacity for that time}} \times 100$$

and is expressed as a percentage.

A nuclear reactor which runs continuously for long periods of time could have a capacity factor close to 80 per cent although most do not achieve this. The Vestas 27 wind turbine installed in Wellington, which has a power rating of 225 kW, produced its first 1 000 000 kWh in 366 days. This converts to a capacity factor of $\dfrac{1000000}{225 \times 24 \times 366} = 50.6$ per cent

which is remarkably high for a wind turbine. In many countries such as Denmark, capacity factors in the order of 20 per cent are economically acceptable.

The Clyde dam, which has a rated power of 432 MW at present, generated 2793 GWh in fifteen months (454 days, 31/3/93 to 30/6/94) which represents a capacity factor of

$$\frac{2793 \times 1000}{432 \times 454 \times 24} = 59.3 \text{ per cent}$$

which is not much higher than the capacity factor of the Vestas V27 wind turbine.

---

the possibility of damage, turbines are shut down when wind speeds exceed the design maximum. For the V27 this speed is 25 ms^{-1} (90 kmh^{-1}).

The Wellington wind turbine, which is able to supply about 100 homes when operating at rated capacity, until recently held the world record for electrical generation for a machine of its type. It has generated 3.6 TJ (1 GWh) in its first year of operation, a capacity factor of greater than 50 per cent.

Larger machines with rated outputs of up to 4 MW have been constructed, but it has proved more practical to operate 'farms' composed of groups of smaller turbines in the 75-500 kW range. Machines of this size are more reliable in stormy conditions, and cheaper to construct. Most grid-connected generators presently being developed are in the range of 300 kW to 1 MW rated capacity.

From New Zealand's point of view, electrical generation from wind turbines has many advantages. The country has a large number of sites where annual average wind velocities of between 7 and 12 ms^{-1}

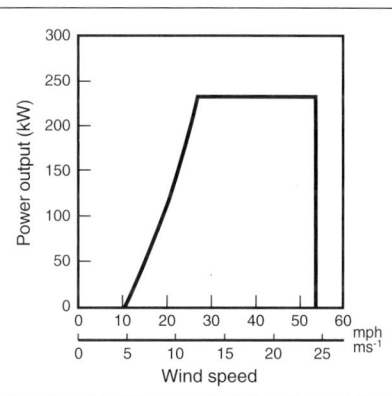

Figure 5.2
The Wellington Vestas V27 wind turbine: rated power and wind speed.

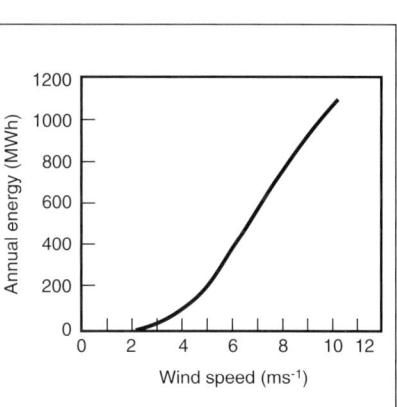

Figure 5.3
Annual energy output for a 225 kW wind turbine, like that in Wellington, related to wind speed.
(Energy Efficiency and Conservation Authority)

The Energy of the Wind

Wind energy is harnessed by wind turbines sited either alone or in wind farms. The function of the turbine is to convert the kinetic energy of the wind into electrical energy.

The kinetic energy of the wind is represented by the familiar expression

$$E_k = \tfrac{1}{2}\, mv^2$$

and the energy available each second (power) is equal to the kinetic energy of the mass of air which travels past the turbine blades every second.

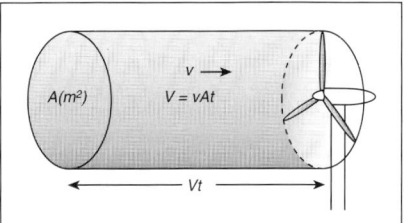

If the turbine blades sweep out an area A, a cylinder of air, which has a cross sectional area of A and is travelling at a velocity v, will pass the blades. Over time t the cylinder will have a length vt, and a volume vAt.

Now density is defined as mass per unit volume.

$$\rho = \frac{m}{v} \quad \text{which rearranges to give} \quad m = \rho V$$

So, replacing m by ρV, the kinetic energy of this air is

$$E_k = \tfrac{1}{2}\rho V . v^2$$

but we have already shown that the volume V of the cylinder of air which passes the blades in time t is vAt, and so the kinetic energy possessed by this cylinder of air is

$$E_k = \tfrac{1}{2}\rho V . v^2 t = \tfrac{1}{2}\rho v^3 At$$

So the power available to the turbine blades is:

$$\frac{E_k}{t} = \tfrac{1}{2}\rho A v^3$$

We see that it increases as the cube of the velocity of the wind.

power is proportional to wind velocity [3]

As a sample calculation, the Vestas V27 turbine in Wellington is rated at 225 kW, which it develops at a wind speed of 12 ms⁻¹.

The radius of the swept area of the Vestas' turbine blade = 13.5m and so the swept area

$$A = \pi \times 13.5^2 \,(\text{m}^2) = 573 \,(\text{m}^2)$$

At sea level the density of air is 1.2 kgm⁻³, and this means that the total power available is

$$0.5 \times 1.2 \times 12^3 \times 573 = 594 \text{ kW}$$

Notice that although 594 kW of power is available to be extracted by the blades, the rated power of the turbine is 225 kW, giving a conversion efficiency from wind to electrical energy of 38 per cent. There are several reasons why the efficiency of a wind turbine cannot approach 100 per cent. First, as the blades rotate, the tip of the blade creates a vortex, much like the vortex from an aircraft wing. To give some idea of the energy loss to this cause, half the drag on a commercial aircraft such as a Boeing 747 is drag due to wing vortices. The power of two of the 747 engines is used just in overcoming tip drag. Further drag and turbulence occurs elsewhere on the blade.

A further point is that for the turbine to achieve 100 per cent efficiency would mean that all the kinetic energy has been extracted from the air. In other words, the air has stopped! This is clearly impossible.

To summarise, the maximum percentage of kinetic energy that can be transferred to the blades is known as the Betz Limit. It is almost 60 per cent. It should be stressed that increasing the number of blades of a turbine makes no difference to this situation.

The classic farm windmill is less efficient than the modern wind turbine designs. The Wellington turbine achieves up to 38 per cent efficiency in its conversion of wind energy to electrical energy. There are energy losses in the gearbox and electrical generator, as well as in the actual blade performance at the rated wind speed. When the wind is above the speed at which the turbine just achieves its rated power, which is called its rated speed, the efficiency decreases, as the electrical generator cannot deliver more than 225 kW. Choosing the best value for the rated wind speed is called 'optimising' the turbine.

Potential wind farm sites	Wind speed [m/s]	Energy output per farm [GWh]
1. Far North	7.4	1074
2. West Coast Auckland	7.2	393
3. Coromandel/Kaimai Ranges	8.8	1507
4. Cape Egmont/ Taranaki Coast	8.4	866
5. Manawatu Gorge	10.5	1945
6. Wellington Hills and Coast	10	223
7. Wairarapa	8.7	1152
8. Marlborough Sounds Hills	7.8	1064
9. Banks Peninsula	8	461
10. Canterbury River Gorges	6.4	518
11. Inland Otago	6.4	1284
12. Foveaux Strait and SE Hills	8	2171
Total energy in GWhy⁻¹/year		**12656**

Figure 5.4 A conservative estimate of the energy potential from wind farms. (Ministry of Commerce)

Figure 5.5 Location of possible wind farms. (Ministry of Commerce)

are reliable over extended periods of time (Fig. 5.4). And, as Fig. 5.3 shows, the annual energy output of the wind turbine is very sensitive to wind speed, the power available being proportional to the cube of the wind speed.

Wind farms at these sites could generate economically at least 45 PJ of energy each year, which is almost one half of New Zealand's electrical demand. In terms of what is economic overseas, the total potential is enormous, being as high as 225 PJ (50 000 GWh) or over twice our present demand. Many of these sites are sufficiently remote from habitation to avoid visual or noise impact, or interference with television, yet they are close enough to existing high voltage transmission lines and to demand to avoid further capital cost of grid construction (Fig. 5.5).

Wind farms can be built in a modular fashion, in step with the changing electrical demands of our society. The farm can even be removed at a later date, and while in operation over 90 per cent of the land devoted to it can still be used for agriculture.

However, wind energy is not a constant energy source and problems of supply stability may arise if more than 20 per cent of the grid capacity is supplied by the wind. On the other hand, although the wind in

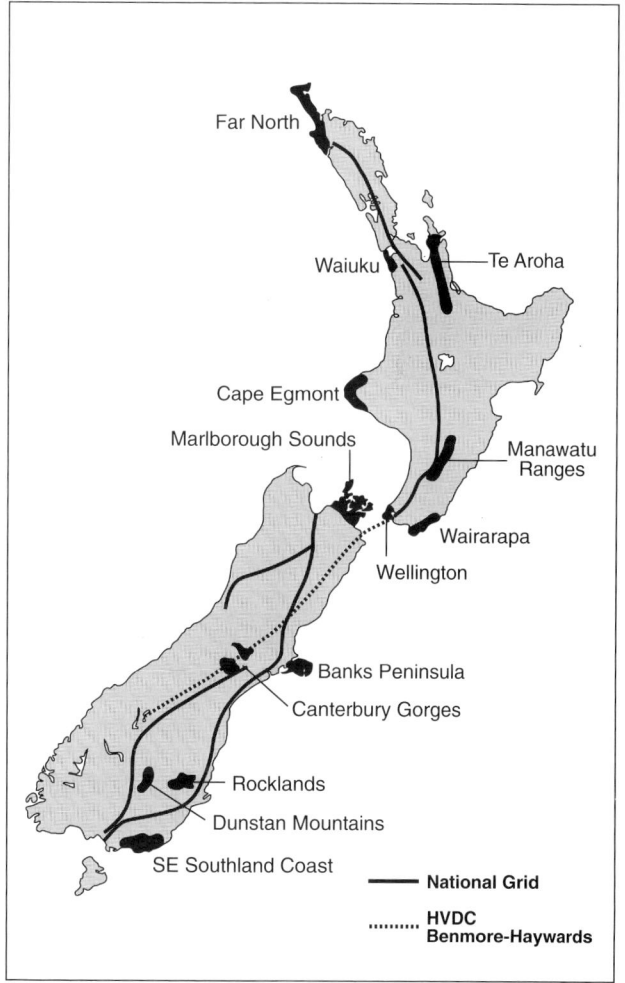

Far North

Waiuku — Te Aroha

Cape Egmont

Marlborough Sounds — Manawatu Ranges

Wairarapa

Wellington

Banks Peninsula

Canterbury Gorges

Rocklands

Dunstan Mountains

SE Southland Coast

—— **National Grid**

.... **HVDC Benmore-Haywards**

The Noise Impact of Wind Turbines

Our perception of sound is caused by variation of air pressure detected by the human ear. The frequency range of detection is from about 20 Hz to 16 kHz, although the upper limit, which might be as high as 20 kHz for a young person, decreases with age. Noise is defined as unwanted sound.

The human ear can detect a pressure ratio from the weakest (quietest) to the strongest (loudest) sound of over a million. This would make a linear scale unwieldy and so we use a logarithmic scale, which has the units of decibels (dB). For sound pressure we take the zero as being the threshold of hearing (0 dB) and on this scale a car passing at 100 m develops a sound level of 50 dB which is 10^5 times the sound pressure at the threshold of hearing. Doubling the sound power of the source increases the sound level at a particular position by 3 dB.

The human ear is not equally sensitive to sounds at all frequencies, being progressively less sensitive to sounds as the frequencies go below 1 kHz or above 3 kHz (Fig. 5.6).

When measurements are taken, the sound is often filtered to a response similar to the human ear and recorded in units of dB(A). This means that when industrial noises, such as from factories, traffic or wind farms, are measured they are reported as dB(A).

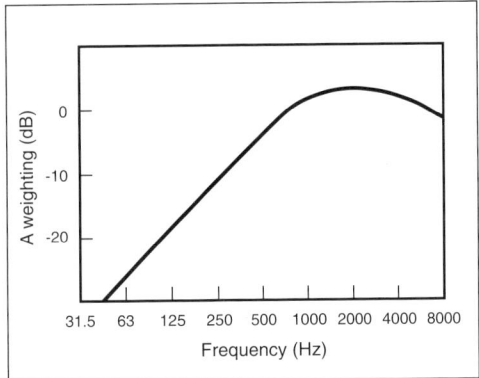

Figure 5.6 Measuring dB(A).

At present, experience from wind farms indicates that during average wind conditions (5-10 ms^{-1}) sound pressure levels at 300 m are of the order 45 dB(A), which is only just above night-time back-ground levels (40 dB(A)) and well below noise levels 100 m from a highway (Fig. 5.7). Indeed, one of the authors stood in a wind farm 350 m from a busy British main road and could not hear the turbines because of traffic noise.

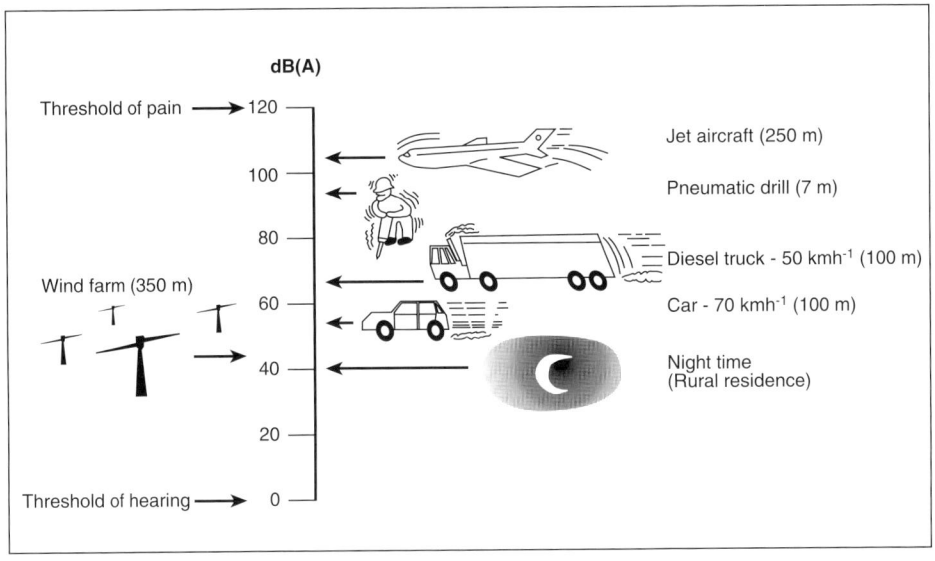

Figure 5.7 Common sound levels measured in dB(A).

New Zealand is rather variable on a daily basis, it is surprisingly consistent on a monthly or an annual basis.

In the New Zealand situation, to supply 10 per cent of the 100 PJ of electrical energy generated annually would require 3000 turbines similar to the Vestas V27. If these were arranged in farms of 100 turbines each, located around the country, there would be thirty farms. As the current capital cost of a wind turbine of this size is about $600,000, the total cost of this installation would be close to $2 billion, if the cost of connection to the grid is ignored. As this is little more than the capital cost of a further major dam, which would also require major extensions to the grid if constructed, it would seem that the erection of wind farms would be not only the more flexible choice, but also the more cost-effective.

Our electrical energy demand continues to increase and at present 70 per cent of the demand is met by hydro. Considered in detail, the construction of a further 350 MW dam proposed for Tuapeka Mouth on the Clutha River can be compared with the equivalent number of wind farms; if we assume capacity factors of the dam and wind turbines were similar, ten farms each containing 100 turbines rated at 350 kW would be required.

Hydro dams require massive capital investment. The Clyde Dam has cost at least $1.7 billion, of which $337 million was for land stabilisation around Lake Dunstan. It is suggested that the Tuapeka Mouth dam could cost a similar sum. Large hydro dams flood agricultural land and often require the resiting of roads and population. This would happen in the case of the Tuapeka Mouth dam, and ECNZ is meeting considerable local opposition over this point.

The power supplied by a dam is often more than is required at the time of construction, although due to extended construction periods, and occasionally to marketing, demand eventually rises to meet the supply. In terms of employment, a massive labour force is assembled during the construction of a large dam, but this force is disbanded as the dam is completed, unless a start can be made on another dam. Major dams can only be located where there are large flows of water, and where large storage lakes can be formed. These places are often of scenic importance and usually are located far from the demand for power. In contrast, wind farms may be sited at any location where the wind is suitable. They do not flood farmland, which means that farms do not have to be purchased, and the land remains productive throughout the life of the wind farm, reducing both the capital and the running cost of the energy generation. Finally, wind turbine manufacture, erection and maintenance has the potential to provide employment for a good number of skilled workers on a continuing basis.

At the present time, wind energy is commercially competitive with large scale hydro schemes. At $2000 per installed kW, the capital cost of a wind farm is similar to that of a dam. The Clyde Dam, which in 1995 has a generating capacity of 432 MW, cost almost $4000 (kW)$^{-1}$, although this cost would be reduced to $2800 (kW)$^{-1}$ if the dam's output were increased to its designed capacity of 600 MW. Here it must be pointed out that even if the peak power output of the Clyde Dam were to be increased, its annual energy output, which depends on the total amount of water available, might not increase. In other words, the dam's capacity factor might fall. It might also be argued that the life of a dam is longer than the expected life (about twenty years) of a wind turbine, and that the capital cost of power over the lifetime of a dam is lower than that for a wind farm, but this ignores the need to regularly refurbish the generating machinery of the dam. Wind turbines and dams have similar refurbishment needs and the turbine towers can be as permanent as the concrete dams.

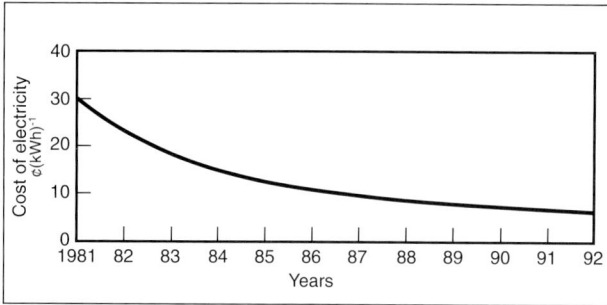

Cost of electricity c(kWh)⁻¹ (y-axis): 0, 10, 20, 30, 40
Years (x-axis): 1981 82 83 84 85 86 87 88 89 90 91 92

Figure 5.8
The decreasing cost of electricity generated by wind turbines. (ECNZ)

The cost of generation of the Clyde Dam has been estimated to be between 12-17c $(kWh)^{-1}$. The generating cost for new dams is certainly much higher than for existing dams and is unlikely to be less than the 12c to 17c range. ECNZ estimates the cost of wind energy might range from 6c to 12c kWh^{-1}, and in view of their experience with the Wellington turbine, a value in the lower part of the range might be expected. It is certain that the inflation-adjusted cost of wind generation is decreasing (Fig. 5.8) whereas that of hydro is at best static, and is most likely to increase, as the cheapest sites in this country have already been developed.

If New Zealand generated electricity from a mix of hydro and wind we would be less exposed to difficulties during periods of low rainfall. If wind energy was used when available, hydro storage lakes would become less depleted and represent stored energy. In this manner, power shortages such as occurred in the winter of 1992 might be avoided. Even if the 'capacity factor' of a wind farm is less than that of a hydro dam, the mix would be suitable for New Zealand's conditions. New Zealand would be able to expand its generating capacity as required, and to guard against conditions when hydro power is not available, without increasing our use of fossil fuels.

Remote Area Power Schemes (RAPS) might also benefit from the introduction of wind energy. At present islands such as the Chathams, Stewart and Great Barrier rely on diesel generation for their electrical energy. This is particularly costly, for example Stewart Islanders are paying up to 75c $(kWh)^{-1}$. A model worth consideration is the system using a mixture of diesel, wind and battery storage which has been installed on Rathlin Island, off the coast of Northern Ireland.

Rathlin Island Resource

The scheme is based on three 33 kW wind turbines and three diesel sets with rated powers of 48, 80 and 132 kW respectively. A battery system of 110 batteries of 250 Ah capacity operates at 220 V DC and can store 3.6 GJ (100 kWh) of electrical energy. On Rathlin Island, the 'extraordinary' wind conditions of 10 ms⁻¹ yearly mean velocity (not uncommon in New Zealand!) mean that the diesel sets are shut down for over 80 per cent of the time. During 1993, its first full year of operation, the scheme generated 250 000 kWh, of which wind energy supplied about two-thirds.

Such schemes are beginning to be operated in remote areas of New Zealand. On Great Barrier Island, both the police station and the health centre are supplied with electricity by wind turbines, and in North Canterbury a farm which is several kilometres from the grid supply derives all of its electricity from a wind turbine, with a battery back-up from sixty two-volt cells which store enough energy for a week.

Delabole Wind Farm – A Case Study

In 1980 a proposal was made to construct a nuclear power station in Cornwall, a scenic region of the United Kingdom. This was vigorously opposed by many local people. One family, the Edwards family of Delabole, decided to investigate an alternative in the form of a wind farm. Planning permission for the farm was obtained in August 1991, and a company, Windelectric Ltd, set up to construct the NZ$8 million wind farm. Work for the foundations of the turbine towers was started on 30 August 1991.

The Delabole Wind Farm became operational in December 1991. The construction of the farm took three months, but the erection and commissioning of the ten 400 kW turbines took only fifteen days. The turbines are all situated in the existing hedgelines and as there are no access roads, and all the cables are underground, the farmland is totally undisturbed. The turbines themselves are serviced every six months, just like a car. In their first three years of operation there have been no major problems.

Power is generated at 690 V and is transmitted down a cable within the tower to a 690/11 kV transformer in a hut at the base of each tower. From here it enters the grid. The ten turbines have an annual output of 12 million kWh, which is the average electrical energy consumption of 3000 average homes (a small town). The capacity factor of of the wind farm is 30 per cent.

To produce the same quantity of energy from fossil fuels would mean the combustion of 2000 tonnes of oil or 5000 tonnes of coal and would result in the production of 12 000 tonnes of carbon dioxide and 120 tonnes of sulphur and nitrogen oxides.

Local reaction to the wind farm has been very positive, so much so that the Delabole Wind Farm visitors' centre has received more than 160 000 visitors in three years. It is something of a tourist attraction!

Delabole Wind Farm: blade from turbine (top) and the farm in operation.

6 Solar Energy

Energy equivalent to over 10 000 times human society's annual consumption and 100 times our fossil fuel reserves reaches the Earth from the Sun every year. On a sunny day, the total solar radiation received per square metre of horizontal surface in New Zealand varies from 24 MJ for Northland during January to 3.6 MJ for Southland during June. Averaged over New Zealand, the daily received solar energy would be 15 ($\pm$1.5) MJ m$^{-2,}$ which is a little less than the energy of half a litre of gasoline (energy content of gasoline 34 MJ litre^{-1}). To stress the point, the total solar energy falling on the roofs of New Zealand houses amounts to 600 PJ annually, which is more than New Zealand's annual primary energy demand.

Some areas of New Zealand are particularly suited for the use of solar energy. Blenheim, Nelson-Motueka and the Bay of Plenty experience an average of 2350 hours of bright sunshine each year, but even Southland and coastal Otago (which lie on the northern fringes of a belt of cloudiness) will receive 1700 hy^{-1}. It was estimated by Bensemann in 1965 that using the simplest technology, a passive water heater, an electrical energy saving of 1700 kWh y^{-1} (~$170) could be made in a dwelling housing three or four occupants and requiring 200 litres of hot water every day. Despite this, only 6500 of the one million plus households in New Zealand had solar water heaters by 1982; this situation has hardly improved to the present day. From the current data available, there are probably no more than 10 000 domestic solar water heaters installed in New Zealand. There is clearly an enormous potential for the use of this 'wasted' energy.

Use of solar energy does not contribute to any form of pollution, nor to greenhouse gas emission, and it would be difficult to claim that it has a major visual impact. Indeed, solar energy collection presents an opportunity for individual households to contribute to energy production, moving each dwelling towards 'self sufficiency'.

There are several methods by which solar energy might be utilised. At present the proven technologies are: active and passive space and water heating, the direct production of electrical energy from photovoltaic (PV) cells, and from solar thermal units.

Although it has already been mentioned that only a small percentage (about 1 per cent) of New Zealand's homes have been provided with means for solar water heating, even so we have made use of passive solar heating for many decades. By building houses with large north-facing windows, New Zealanders have been able to take advantage of sunlight to warm their homes. Although specific figures do not appear to exist for New Zealand conditions, the United Nations have estimated that buildings use 50 per cent of the world's energy demand (Fig. 6.1): 5 per cent in their construction and 45 per cent in heating, cooling, lighting and ventilation. It is clear that improvements in the energy efficiency of buildings would make a significant reduction in our energy demand.

Modern 'passive solar' architecture uses

Figure 6.1
Global energy use.
(United Nations)

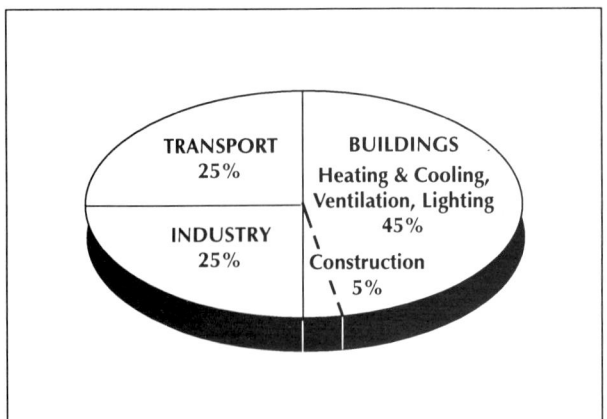

Natural lighting and ventilation is used as much as possible at the Nelson Library. (Photo: courtesy David Wallace)

the building as a solar energy collector and storage system, providing a comfortable temperature and lighting level for its occupants. By the use of large windows, well insulated walls and massive concrete floors which warm up during the day and release heat at night, day-long comfortable temperatures may be provided. Cooling may also be achieved by carefully designing the home to allow convection currents to flow through the building. Shading will allow maximum solar trapping in the winter when the sun is low, and maximum shielding in the summer.

Modern architecture is moving to 'active solar', where the use of heat pipes, electronic controls, modern glass and building materials allows better heat distribution. It is likely that in just a few years we will see 'intelligent' houses, in which micro-processors control 'active solar' techniques to maintain a pleasant living temperature throughout the day using only solar energy. Similarly, large office buildings are increasingly lit by natural sunlight distributed around the building by 'light pipes'. As 30-50 per cent of the energy

used in commercial buildings is for lighting, the use of natural lighting will save energy. The Nelson Library building is an example of what can be done in this respect. This building uses natural sunlight as much as possible, and its natural ventilation means that the building does not have an air conditioning system. These two innovations have meant a capital saving of $40,000. It is expected that annual energy savings of $4000 will also be achieved.

About 20 000 homes are built in New Zealand each year. As the 'average' household uses about 200 litres of hot water every day, even if each of these new homes were fitted with solar water heating a national energy saving of 120 TJ (34 GWh) would be expected. If a continuous retrofitting programme of older buildings were undertaken, the savings could eventually amount to 6 PJ y^{-1}, which is one fifteenth of the New Zealand electrical energy consumption.

A number of systems have been developed for the generation of electrical energy from solar energy. Solar thermal systems rely on the collection of sunlight by parabolic reflectors, either in the form of troughs which focus sunlight onto pipes running along their axes, or else in the form of mirrors which reflect sunlight onto a central boiler. The reflectors are made to track the Sun at all times by the use of computers.

Commercial parabolic trough systems have been built. In June 1994 a plant with a dish area of 400 m², developed by the Australian National University, began supplying energy to the Canberra supply. The plant generates 50 kW. Studies are proceeding into a proposed 2MW system to be located at Tennant Creek in the Northern Territory. During periods of low sunshine, the plant will use natural gas to ensure continuity of supply of electrical energy. A similar plant has been tested in the Mojave desert for several years. New Zealand conditions are such that this type of plant is not considered to be economic here at present.

The photovoltaic cell is more likely to be of practical use in New Zealand. These

How Photovoltaic Cells Work

When light shines on certain substances, electrons may be emitted from their surface. It was found that for each substance which displayed this effect, the light had to have a frequency higher than some particular minimum. Above this minimum frequency, the electrons which were emitted had energies which increased as the frequency of light increased, in a straight line fashion. It was also found that if the brightness of the light was increased, the energy of the electrons stayed the same, but more were emitted. In other words, the current increased.

Einstein explained this effect in 1905, by using the quantum theory. Although at that time the view that light was an electromagnetic wave had been established, Einstein proposed that it might also behave as a particle. He suggested that the energy (E) of a light particle (photon) was linked to its frequency (f) by the equation

$$E = hf \qquad (1)$$

where h is Planck's constant (6.67 x 10⁻³⁴Js).

The frequency and wavelength of light are linked by the equation

$$c = f\lambda \qquad (2)$$

where λ is the wavelength of the light measured in metres, and c is the velocity of light which is 3.0 x 10⁸ ms⁻¹. The frequency is measured in hertz (Hz).

Einstein suggested that if a photon struck an atom of a substance, it would be absorbed, and its energy would be gained by an electron in that atom. Electrons occupy energy levels in atoms, and if enough energy were to be supplied, an electron might be excited (lifted) from its energy level to the edge of the atom. This process is called ionisation, and the energy required to do this is termed the ionisation energy or work function of that substance. Any excess energy above the ionisation energy would allow the electron to move around an external circuit. In other words, a current would be produced in the external circuit. We call this the photocurrent. The excess energy will be the kinetic energy of the electron, which corresponds to the voltage of the photocurrent. Einstein produced an equation

$$hf = eV + \phi \qquad (3)$$

in which hf is the energy, calculated from equation (1), of the incident photon, eV is the kinetic energy of the electron (e is the electron's charge and V is the voltage of the photocurrent), and ϕ is the ionisation energy of the substance on which the light falls.

When sunlight strikes a semiconductor, the light has sufficient energy to promote electrons of the semiconductor atoms into what is called the conduction band of the semiconductor. The semiconductor atoms losing electrons will become positive, and are considered to be 'positive holes'. In normal circumstances, in a block of semiconductor material, the positive hole would quickly cells, which make use of the photoelectric effect, transform light directly into electricity. They were developed for the space programme, and everybody will be familiar with photographs of the Hubble Space telescope, with its solar panels deployed.

There are two main types of PV cells. Thick film cells are made from crystalline silicon. In their production state they are able to convert up to 17 per cent of the incident sunlight into electrical energy and, although each cell can only produce up to 0.4V, the cells are connected in series and

in parallel into modules. Thick film modules have an expected lifetime of thirty years and it is hoped that efficiencies of up to 36 per cent may be achieved by stacking different cell materials, so that each layer collects one part of the solar spectrum.

Thin film modules have a lower efficiency of 8-10 per cent, and age rapidly because of diffusion inside the cells, giving a maximum module life of ten years at present. They are composed of amorphous silicon, which is vacuum deposited

---

recombine with one of the promoted electrons, and no useful generation of electrical energy would occur.

To make a useful device from, say, a wafer of silicon, one layer of the silicon must be doped with an element such as gallium, which has fewer valence electrons per atom than does silicon. This means that the layer will possess permanent positive holes. We say that the layer is *p*-doped. The other layer is doped with an element such as phosphorus which has more valence electrons per atom than silicon. This latter in *n*-doped.

When light strikes our device, electrons will be excited in the *p*-doped layer. These

electrons would rapidly recombine in the presence of positive holes, but many will migrate to the *n*-doped layer, where because of the *n*-type impurities they will not find positive holes with which to recombine. After a while, the wafer of silicon would reach a state where so many electrons had migrated that they would repel any further migration and equilibrium would be reached. If however an external circuit were provided, so that the electrons could return to the *p*-layer, the process could continue as long as the light shone, and we would have constructed a photocell.

There are limits to the efficiency of this process. Many of the liberated photoelectrons will recombine with positive holes before they can migrate to the *n*-region. This will always be a limitation, but as experience is gained in the manufacture of the materials better diffusion is being achieved. The second limitation is that thin layers of silicon are partially transparent to light. This problem may be overcome by making the surfaces of the cells partially reflective, causing the light to bounce backwards and forwards until it is absorbed. Problems also exist in the connections of the external circuit to the cell. Despite these limitations, Prof. Martin Green's team at the University of New South Wales has produced a cell with a conversion efficiency of 24 per cent.

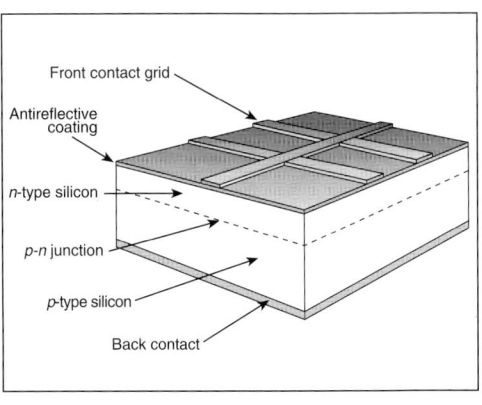

Front contact grid

Antireflective coating

n-type silicon

p-n junction

p-type silicon

Back contact

A schematic drawing of a silicon solar cell.

onto a backing. They are only about 0.01 as thick as thick film cells and so are much lighter and also considerably cheaper. The deposition technique lends itself to mass production, and the cells can be deposited onto a flexible base, making the cells themselves flexible.

The low weight factor of thin film PV modules has enabled them to be used in prototype situations. Between 16 July and 3 September 1990, an aircraft called the 'Sun Seeker' flew 6400 km from California to North Carolina, powered only by PV modules and back-up batteries. Furthermore, the Nissan FEV electric concept car has an array of PV modules on its roof to augment the charging of its batteries. It may not be long before electric cars, which are capable of partially recharging themselves when parked, appear on New Zealand roads.

PV cells have been made from semiconducting materials other than silicon. Some of these materials, such as gallium arsenide, show considerable promise in terms of efficiencies, but at present the technology is very expensive.

Photovoltaic technology has not developed to the state where it is commercially competitive with other methods of electrical generation. The University of New South Wales is at the forefront of solar cell development, and holds the world record for cell efficiency, with efficiencies of 23.5

per cent for thick cells and 15 per cent for thin film cells. As PV cells use no fuel, the cost of electrical production is almost entirely the interest on the capital cost of installation. At the present time, the cost of electrical energy from PV is at least 50c $(kWh)^{-1}$ based on a capital cost of $10,000 $(kW)^{-1}$. However, the leader of the UNSW group, Professor Martin Green, predicts that with increased efficiencies and the benefits of large-scale production, capital costs could become as low as $1000 $(kW)^{-1}$ within ten years, with a corresponding reduction in generation costs.

PV arrays are being used in remote locations. The Great Barrier Island Medical Centre, which was built in 1990, relies for its energy requirements on a PV solar panel which charges a 24V battery bank, and is converted to 240V AC through a DC to AC inverter, as well as a 400 W wind generator, a solar water heater, and a wood-fuelled space heater.

Large PV arrays of up to 200 kW power have been installed and linked to the grid. In Switzerland, a 100 kW array has been placed alongside a motorway, and in Italy a PV array rated at a peak output of 3.3 MW has been completed. Architects have also experimented with PV fascias for buildings. Cladding a building with PV modules is cheaper than cladding it with marble, and the 'space-age' look of the building is attractive. In Switzerland, several office buildings and even a church have been 'PV clad', and the University of Northumbria in northern England has covered the south wall of one of its buildings with a facade of PV collectors, which give a peak output of 16 kW.

New Zealand, located between 47° and 34° S, is closer to the Equator than either Switzerland or the UK. This, together with our less polluted atmosphere, would make the placing of PV modules on buildings particularly favourable in our country.

A large factory, such as that at Oberburg (opposite), might have a roof area in excess of 500m². In Auckland the average

PV array alongside a motorway in Switzerland. (Photo: ENET, Bern 1992)

South wall of a building at the University of Northumbria, in England, covered with PV collectors which give a peak output of 16 kW. (Photo: Andrew Hayward, courtesy BP Solar)

daily energy resource is about 20 MJ m⁻², which means that at 10 per cent efficiency, 1 GJ (3.6 MWh) of electrical energy could be collected by this factory. The energy would be collected in daytime hours, during which the activities of the factory are at a maximum. It is quite conceivable that the factory would be grid-connected, 'exporting' electrical energy during the day and 'importing' it at night.

A house with 50 m² of PV modules could generate 100 MJ (27 kWh) of energy every day, assuming 10 per cent efficiency. This amounts to 36 GJ (10 000 kWh) every year. This is a little more than the 'average' electrical energy requirement of a New Zealand home. If one per cent of the total surface area of Auckland were covered in PV arrays, this would provide 3.6 TJ (100 GWh). Finally, if a little less than 200 km² of New Zealand were provided with PV arrays at their present efficiency and a means of storing the energy were provided, then the country's electricity demand would be met.

In rural areas, power utilities may charge $10,000 or more to connect a house to the domestic supply. If a new house is designed to be energy-efficient, then it may

be cheaper to provide it with PV modules and storage batteries, rather than to connect it to the supply. The Dunedin firm of PCM Solectric has recently equipped a house in Wanaka with a PV array which makes it independent of external electrical supply.

PV research is at present being pursued vigorously, with over $US 6 billion having been invested in research over the last twenty years. The PV development goals of the US Department of Energy are listed in Fig. 6.2.

PV array on the roof of a factory at Oberburg, in Switzerland. (Photo: ENET, Bern 1992)

Elements of a highly efficient power system which is simple to operate and which generates energy from the sun and the wind, developed by a Dunedin company, PCM Solectric Co Ltd. (Photo: Allied Press)

	1991	1995-2000	2010-2030
Module efficiency (per cent)	5 – 15	10 – 20	15 – 25
Electricity price c (kWh)$^{-1}$	50 – 80	20 – 30	10
System lifetime (years)	10 – 15	20	30
Installed capacity (MW)	<50	200 – 1,000	10,000 – 50,000

*Figure 6.2
PV development
goals of the US
Department of
Energy – prices are
in NZ cents,
standardised to
1990 values.*

Besides PV cells using semiconductor materials such as silicon or gallium arsenide, novel cells which mimic the process of photosynthesis are being developed. The Gratzel cell developed by Michael Gratzel is essentially a titanium dioxide layer which supports a compound based on ruthenium. The cell converts sunlight to electrical energy over a wide range of wavelengths. Cell efficiencies of over 10 per cent have been achieved and the cells continue to produce in diffuse light with efficiencies exceeding those of silicon PV cells, giving them an advantage in cloudy areas. A particular feature of the Gratzel

cell is that it is transparent, and so it may be that in the future the windows of large buildings might generate electricity for the building. These cells, which are expected to be commercially available in 1995, might cost as little as $100 (kW)$^{-1}$, which is considerably less than the best predictions that have been made for the eventual cost of silicon PV cells. New Zealand has considerable reserves of titanium dioxide, which is produced as a by-product of steel manufacture from West Coast iron sand. At present the titanium dioxide is used as a paint pigment and paper whitener. In this country, we would be ideally suited to take advantage of this new technology.

Researchers in Italy have synthesised ruthenium and osmium compounds which trap light. They hope that these molecules might be able to mimic the photosynthesis reaction and split water, providing a potentially unlimited supply of hydrogen gas for use as a fuel.

Many people feel that the development of solar energy, particularly in the form of PV cells, offers humanity the best long-term hope for a sustainable energy future.

The Solar Resource

Solar energy results from the processes of continuous nuclear fusion in the Sun. At the core of the Sun, matter is contained at a temperature of 15×10^6 K and a density twelve times that of solid lead. Under these conditions individual atoms cannot exist and there is a 'soup' of protons and electrons. Normally, if two protons approach they repel each other and do not interact, but under the conditions in the Sun's core, if one of the colliding protons is travelling five times as fast as average it will interact with the other proton to form a deuterium nucleus, a positive electron ('positron') and a neutrino,

$$^1_1H + ^1_1H \Rightarrow ^2_1H + ^0_1e + v$$

On average, it will require 14 billion years for this to happen, which is why the Sun has only processed about 4 per cent of its protons in 4.5 billion years.

Once the deuterium nucleus has been formed, it can interact with another proton to form a helium-3 nucleus:

$$^2_1H + ^1_1H \Rightarrow ^3_2He + \gamma$$

A gamma photon is produced during the reaction and this carries away the excess energy. Finally when two helium-3 nuclei collide they form a helium-4 nucleus and two protons:

$$^3_2He + ^3_2He \Rightarrow ^4_2He + 2^1_1p$$

This is the end of the reaction.

The net result is that the Sun, which is 70 per cent hydrogen and 28 per cent helium, is converting hydrogen into helium continuously.

In this reaction some mass is converted into energy according to the Einstein relationship $E = mc^2$.

The Sun is losing four million tonnes of mass, and producing 3.7×10^{26} J of energy every second. Even at that rate, the Sun has only lost 0.02 per cent of its original mass since its formation, and is predicted to shine for another five billion years before its supply of hydrogen begins to run out.

The energy which is initially released in the form of a gamma ray photon is continually absorbed and re-remitted by matter as it moves out from the core. If the photon could fly out at the speed of light it would take 2.5 s to leave the Sun, whereas because of this continual reabsorption and re-emission the energy takes on average ten million years before it leaves.

The Sun has a surface temperature of 5500° C (5800 K). Solar energy arrives at the Earth in the form of electromagnetic radiation across a wide spectrum, which includes not only visible light but extends well into the ultra-violet and infra-red regions. The average intensity at the distance of the Earth's orbit is 1.367 kWm^{-2} and over the whole surface of the Earth this becomes an intercepted power of 174 000 TW. Gases in the atmosphere absorb some of the energies, ozone removing ultra-violet, and water, carbon dioxide and methane removing some infra-red frequencies, and so at sea level, on the latitude of New Zealand, the peak intensity of sunlight is typically 1 kWm^{-2}, or averaged over a 24-hour period, 0.2 kWm^{-2}. Overall, 100 000 TW of energy is received at the Earth's surface, and this drives our environment, allowing for photosynthetic processes which ultimately supply all human food, and also provides the energy for atmospheric circulation.

In other words, the Sun is not only the source of (renewable) solar energy but also of wind, wave and biomass energy systems. In fact, all renewable energy resources, with the exception of geothermal energy which is sourced by radioactive processes within the Earth, and of tidal energy which is obtained from the gravitational potential energy of the Earth/Moon and Earth/Sun system, are derived from solar irradiation.

It must also be recalled that fossil fuels are ultimately derived from solar energy. Our supplies of coal, oil and natural gas were formed from the remains of biological organisms over geological time.

7 Biomass

'Biomass energy' refers to the energy which may be obtained from living material. Plants capture solar energy by the process of photosynthesis, and in New Zealand plants trap and store many times our annual energy demand each year. The flow diagram (Fig. 7.1) illustrates how biomass, readily available as it is, might be converted to suitable energy forms. It represents a convenient energy store in much the same way as does water in a hydro lake, with the added advantages that biomass may be stored for many years if necessary, close to its intended place of use, and that it is easily transported, especially if converted to liquid form.

New Zealanders have always used wood to heat their homes. There are about 200 000 solid fuel stoves and 400 000 open fires in the country, suggesting that up to 60 per cent of homes can be heated with solid fuel. Wood for home heating supplies about 5 per cent of our total energy needs, and a probable demand of 0.6 million tonnes of wood exists for this energy use, which would save 4.8 PJ (1500 GWh) of electrical energy (a further 5 per cent of our present use).

Some countries utilise biomass to a much greater degree. Nepal obtains 95 per cent of its energy requirements from biomass sources, and countries such as Kenya (75 per cent), India (50 per cent), and Brazil (25 per cent) also make considerable use of biomass fuels. In several of the developing countries, the range of biomass fuels extends to the use of dried faeces. In the Shetland Islands, north of Scotland, peat has been the traditional fuel for many centuries and is still used today, as it is in Ireland. Overall, biomass contributes 13-14 per cent of the world's primary energy supply.

New Zealand's climate and land utilisation is very suitable for biomass production. Crops can be grown to produce biomass fuels, and fuels can be obtained from waste materials. In particular, forest arisings, wood processing residues, and straw from grain production are plentiful and are often wasted at present. We might also make use of animal wastes produced in the dairy and pig farming industries.

In order to make greater of biomass, we would need to improve our organisation of crop production, biomass harvest and the transport and storage of biomass material. Biomass has the potential to supply much of our country's energy demands next century, and in particular to replace imported liquid fuels. In this discussion of biomass energy it is convenient to discuss the utilisation of wastes separately from the use of deliberately planted materials ('energy cropping').

Possibly New Zealand's greatest potential for biomass energy production will be in the use of forest arisings, which are the waste materials produced in the felling and removal of trees from forestry plantations, and in the use of waste produced in wood processing. Besides the sale of these materials as domestic firewood to homeowners, there is a huge potential for the production of either electrical energy or of liquid fuel from wood.

New Zealand's present annual wood harvest is 4 million tonnes, measured as oven-dry wood. This is expected to rise to 10 million tonnes by the year 2000 and to reach 13 million tonnes by 2010. Fig. 7.2 shows the predicted expansion of our forestry industry to 2012. Column 3 of this table, 'Area unsuitable for whole tree harvesting', indicates the percentage area of each region which might be used for energy cropping. In the harvesting of forestry there is inevitably waste, consisting of defective logs, and the branches and upper parts of the tree, which are not sold at present. By 2010, there could be 6 million tonnes of dry waste wood available for energy production. This wood could be used for heat production. Industry cur-

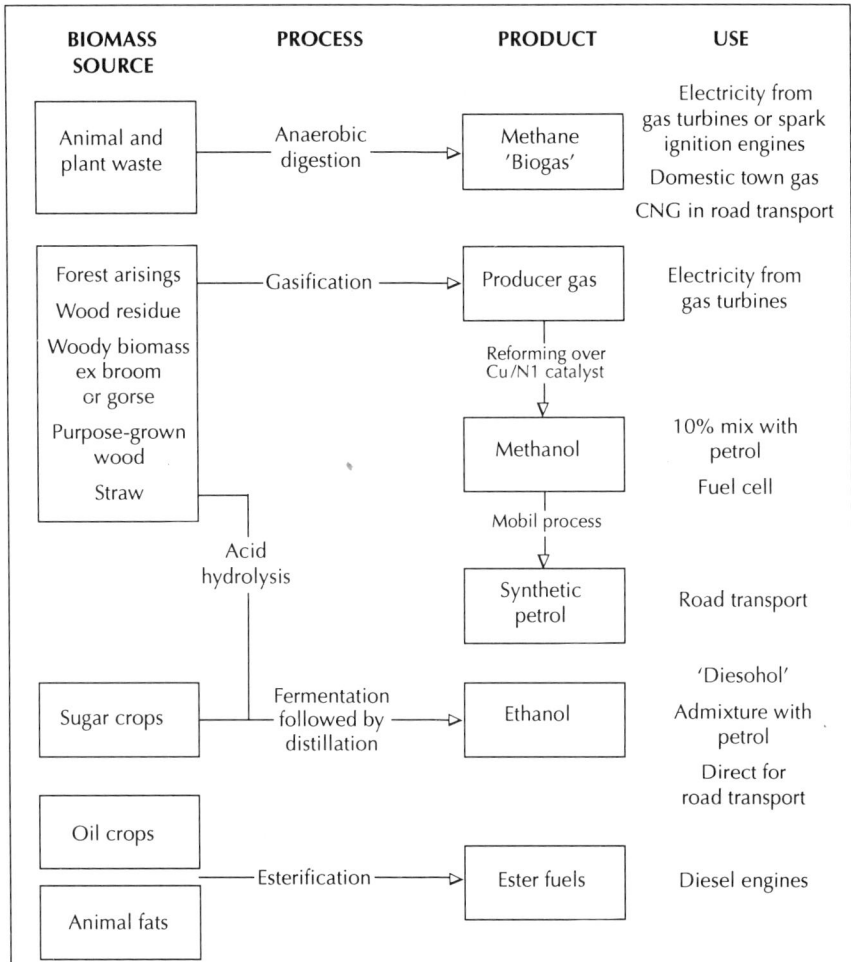

Figure 7.1
How biomass
might be converted
to suitable energy
forms.

BIOMASS SOURCE — **PROCESS** — **PRODUCT** — **USE**

| Animal and plant waste | Anaerobic digestion | Methane 'Biogas' | Electricity from gas turbines or spark ignition engines / Domestic town gas / CNG in road transport |

Forest arisings / Wood residue / Woody biomass ex broom or gorse / Purpose-grown wood / Straw — Gasification → Producer gas — Electricity from gas turbines

Reforming over Cu/N1 catalyst ↓

Methanol — 10% mix with petrol / Fuel cell

Acid hydrolysis

Mobil process ↓

Synthetic petrol — Road transport

Sugar crops — Fermentation followed by distillation → Ethanol — 'Diesohol' / Admixture with petrol / Direct for road transport

Oil crops / Animal fats — Esterification → Ester fuels — Diesel engines

Figure 7.2
New Zealand's
biomass potential
analysed by
selected regions.
(Ministry of
Commerce)

Region	Stocked forest area (ha x 10³)	Area unsuitable for whole tree harvesting (%)	Predicted future merchantable roundwood (m³/annum x 10³)				Potential area for energy plantations (ha)*
			1992	1997	2002	2012	2012
Bay of Plenty/Taupo	438	50	6,550	7,160	8,305	9,505	62,700
East Cape	82	40	110	335	705	1,670	17,250
Hawkes Bay	67	25	420	745	1,740	1,760	42,450
Wairarapa/Wellington	35	40	470	560	725	830	20,250
Nelson	87	40	815	1,350	1,770	1,915	7,950
Marlborough	45	40	135	320	680	825	12,150
Canterbury	53	40	325	385	635	865	69,900
Otago	74	40	475	675	1,285	1,530	71,250
New Zealand	1,239	40	11,390	14,200	20,310	24,820	65,600

* Based on 15% of available 'lucerne 1 and 2' land

rently uses 1.2 million tonnes of coal each year. At present, except for in locations close to forestry or wood processing sites, wood cannot compete with coal on a cost basis. However, the introduction of a carbon tax or an accounting method that took account of environmental factors could alter that situation.

If the wood, either as arisings or as processing waste, were to be chipped, it would represent an easily transportable fuel with an energy content 20-40 per cent that of oil, depending on its water content. This chipped wood could be converted to a gas to be burnt in a gas turbine, generating electrical energy. A Swedish power plant gasifies wood waste under pressure and uses the process gas to drive a Rolls Royce RB 211 gas turbine, with an overall fuel-wood to electricity efficiency of 40-45 per cent. It is estimated that the capital cost of such a plant would be $800 $(kWh)^{-1}$. Such power plants would have an added bonus, in that they remove the need to dispose of wastes.

Certainly one might expect wood processing factories to become energy self-sufficient by this means. It is by no means inconceivable that permanent or semi-transportable power plants will one day convert waste wood into electrical energy, to be used either locally or to be supplied to the national grid.

A step beyond the utilisation of wood waste involves the cropping of vegetation for power production. Trees can be grown specifically for fuel-wood and cut repeatedly (coppiced) to provide a supply of biomass energy with a harvest yield of up to 370 GJ $(ha y)^{-1}$. Several species of trees can be cut up to five times before replanting is needed, and their foliage can be used for animal feed during this time.

Southpower of Christchurch have suggested the cropping of broom (*Cytisus*), a noxious weed. It has been calculated that electrical energy could be generated from this material at a cost of 3.7c $(kWh)^{-1}$. The biomass could be stored dry for use in periods when hydro lake storage is low.

Chemical production from wood would be most useful in the New Zealand context. With current technology it would be possible to produce 600 kg of methanol per tonne of dry wood, as explained in the box opposite. To accomplish this, wood is first gasified using oxygen to eliminate dilution of the products with nitrogen. The hot gases are cooled to 350° C and after removal of carbon dioxide are passed over a copper-nickel catalyst in which a reaction between carbon monoxide and hydrogen produces methanol.

$$CO + 2H_2 \Rightarrow CH_3OH$$

This methanol could be used as a chemical feed stock, or converted via the Mobil process to petrol

$$n\,CH_3OH \quad \overset{\text{zeolite catalyst}}{\Rightarrow} (CH_2)_n + nH_2O$$

Alternatively the methanol could be mixed in 10-15 per cent proportion with petrol, or used directly in a fuel cell. We discuss this further in Chapter 12. Methanol does have a disadvantage as a liquid fuel in that it is highly poisonous.

An alternative conversion of wood to ethanol could produce 200 kg of ethanol for every tonne of wood. The wood is first treated with water in the presence of a sulphuric acid catalyst, to break the cellulose into simple sugars. This is accomplished by reacting the wood with superheated dilute acid at 170-200° for three hours.

$$(C_{12}H_{20}O_{10})_n + 2n\,H_2O \Rightarrow 2n\,C_6H_{12}O_6$$

The sugars are then recovered and fermented with yeast,

$$C_6H_{12}O_6 \Rightarrow 2C_5H_5OH + 2CO_2$$

and the alcohol purified by distillation. Ethanol has an important use as an octane enhancer in petrol. The blending of up to 5 per cent ethanol improves the octane rating of lead-free petrol. It may also be used directly as a road fuel. Brazil has four million cars running on pure ethanol, which is produced by fermenting sugar from sugar cane, and another million which burn petrol containing 20 per cent ethanol.

The production of ethanol generates 700 000 jobs within the Brazilian economy.

In Australia, 'diesohol', a blend of diesel fuel with up to 15 per cent ethanol, together with an emulsifier, has been trialled in three buses running 80 000 km y⁻¹ each in Canberra. The fuel can be used in unmodified diesel engines, and the alcohol component of the fuel, which is made from waste starch by fermentation, does not increase the CO_2 in the atmosphere.

It is not expected that ethanol will become economic as a fuel in this country, unless oil prices exceed $50 per barrel. Nevertheless, there are reasons to pursue the technology. In the Brazilian case, the production of ethanol is made worthwhile by the enormous amount of waste material resulting from their annual sugar production of twenty-five million tonnes. Although ethanol is at present more expensive than oil, of which Brazilians have reserves, nevertheless it is a clean-burning fuel, which means that Sao Paulo, a city of twenty million inhabitants, suffers less urban air pollution than many smaller cities, where petrol is the dominant fuel.

It is possible to grow crops specifically to produce liquid fuels. One such crop, rape (*Brassia napus*), yields an oil (rapeseed oil), which when converted into its methyl ester gives rape methyl ester (RME) which can be used directly in diesel engines. A tonne of crushed rapeseed produces 320 kg of oil with an energy content of 12 TJ of energy. RME has been used successfully to fuel buses, lorries, tractors and taxis, and except for its cost, which is about twice that of diesel at present, it has proved equal to diesel in its performance as a fuel. The French have announced plans to utilise 700 000 hectares of land for RME production.

The sap of a Brazilian tree, *Cabaifera Langsderfii*, can also be used, in an unprocessed state, to fuel diesel engines. South Africa has utilised sunflower oil for that purpose.

Research into the production of liquid fuels from algae has been conducted in the

Wood Gasifiers

A gasifier consists of a sealed reaction vessel into which wood may be dropped from the top and air injected at the bottom.

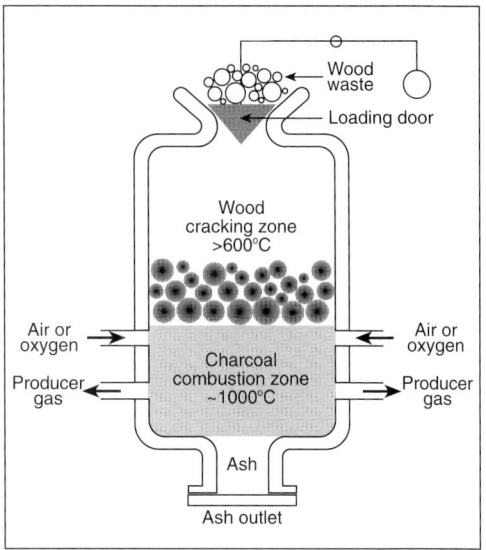

Cross section of a conventional wood gasifier.

At the base of the reactor is a layer of charcoal at least 0.3 m thick. The charcoal reacts with oxygen from the air, to produce carbon dioxide and heat.

$$C + O_2 \Rightarrow CO_2 \quad \Delta°H = -394 \text{ kJ mole}^{-1}$$

The hot gases from this combustion region cause the wood in the upper part of the vessel to decompose to charcoal and a complex mixture of gases. As oxygen has been removed by reaction with charcoal lower in the reactor, no combustion occurs in this region.

Provided the temperature in this region is in excess of 600° C, complex molecules such as tars and hydrocarbons are broken down (cracked) into simple molecules.

Most of the charcoal reacts with carbon dioxide and steam from lower levels in the reactor to produce carbon monoxide and hydrogen in reactions which are endothermic:

$$C + CO_2 \Rightarrow 2CO \quad \Delta°H = +172 \text{ kJ mole}^{-1}$$

$$C + H_2O \Rightarrow CO + H_2 \quad \Delta°H = +131 \text{ kJ mole}^{-1}$$

The overall products of gasification are complex, but include carbon monoxide (14-40 per cent) and hydrogen (6-42 per cent), depending on the design of the reactor and whether air (lower percentages) or oxygen (higher percentages) are used.

These gases may be used as the feed stock for chemical processes or as a fuel, the products from 3.0 kg of wood having the potential to replace one litre of diesel.

US, and claims have been made that a pond with a surface area of one hectare could produce 100 000 litres of fuel every year from the cropping of algae. This represents twice the yield per hectare that could be obtained by growing rape.

The anaerobic fermentation of organic material produces gases which contain high proportions of methane mixed with 30-40 per cent carbon dioxide. Such gas is termed 'biogas'. During the past twenty years experiments at Invermay, near Dunedin, have established the feasibility of biogas production from purpose grown crops. Biogas is also produced in large quantities at sewage works, landfill sites and meat-works.

In a typical biogas facility, waste in the form of a slurry is fed into a large fermentor tank where it is maintained at about 35°C while being stirred. The biogas is collected and may be stored either in large butyl bags, steel gas holders or as compressed gas. The gas produced is similar to CNG in its fuel properties and can be used to fuel vehicles or generate electricity.

BMW's car which runs on biogas. Photo: BMW AG)

Can We Grow Our Own?

If we assume that the average car uses about 1250 litres of petrol each year (costing about $25 a week) and that there are 1.6 million petrol-powered cars in New Zealand, we can estimate that private motorists require 2.0×10^9 litres of petrol annually. (In 1992 the total consumption of fuel by private motorists was this figure, when CNG and LPG was taken into account.)

A hectare of fodder beet could produce about 60 tonnes of carbohydrate each year, and 28 tonnes or 40 000 litres of liquid fuel could be made from this amount of beet. So to grow our own liquid road fuel for private motoring we would need a land area of

$$\frac{2 \times 10^9}{4 \times 10^4} \text{ hectares}$$

which is 5×10^4 hectares or 500 km². This is less than the area of Lake Taupo (600 km²).

It can be seen that biomass could easily provide both our commercial and non-commercial liquid transport fuels, without tying up too much of our productive agricultural land.

Many major industries in New Zealand run biogasification plants. Not only does biogas represent a significant energy source, but it is inevitably produced in the treatment of biowaste. As the capital cost of biogas production is acceptable, being between $2500 and $5000 (kW)⁻¹, depending on the scale of the plant, biogas could generate 1-2 PJ of energy if we treated existing wastes. It has been calculated that this energy would cost between 2 and 5c (kWh)⁻¹, which would make the process profitable for large plants.

Biogas is also produced in landfill sites where municipal solid waste (MSW) is dumped. Typically, 60-70 per cent of MSW is organic in nature, including paper, wood, garden wastes and food scraps. In the conditions of the landfill site, anaerobic fermentation will occur to produce a gas which contains 40-60 per cent methane mixed with carbon dioxide. Generally landfill sites are chosen to be sited on impervious materials such as clay, in order to reduce the leaching of water. Some sites have even been lined with polythene sheeting for this purpose. This limits the movement of gas, which tends to become trapped at the site. When the gas does find its way to the atmosphere, it causes destruction of vegetation due to oxygen depletion, and often, unpleasant smells or even fires or explosions. If pipes are laid in or later driven into the landfill, gas may be released, either under its own pressure or else by pumping. The gas often needs to be treated to remove acidic impurities, such as hydrogen sulphide, as well as the considerable quantities of carbon dioxide that it contains. It may then be used as a fuel for heat or electrical generation, or else for the vehicles of the landfill site.

In 1990 it was estimated that the total methane emission from New Zealand landfill sites was in the order of 100 000 tonnes each year, with the five main centres contributing 50 000 tonnes. In 1994, citizens threw away, on average, between 500 and 750 kg of waste annually. Refuse disposal is expected to increase, and so the production of landfill gas must also be expected to increase. At present two landfill sites, Greenmount and Rosedale in Auckland, use landfill gas to produce 234 TJ (65 GWh) of electrical energy each year, and Energy Direct in Wellington is developing its Silverstream site. Dunedin is constructing a pipeline to reticulate landfill gas, after the removal of carbon dioxide, as the city gas supply. This is expected to be on-stream during 1995. As New Zealand's yearly output of landfill gas might exceed 250 000 tonnes by 2000, and as methane is a more potent greenhouse gas than carbon dioxide, it seems appropriate that we should make efforts to utilise this source. Even where it is uneco-

nomic to generate power, it would be sensible to convert the methane to carbon dioxide by burning the gas as it emerges, in order to reduce our greenhouse emissions.

As landfill sites become filled, it is proving difficult to find suitable new sites close to cities. This is particularly so in countries with higher population densities than our own. In order to deal with municipal wastes, incineration plants have been built near large cities in the Northern Hemisphere. In Europe, Switzerland, Luxembourg and Denmark, countries with little land available for landfill, already recover energy from over 70 per cent of their municipal waste.

The SELCHP (South-East London Combined Heat and Power Company)

Two photographs of EnergyDirect's Silverstream landfill, showing the main gas inlet (top) and the powerhouse. (Photo: Energy Direct)

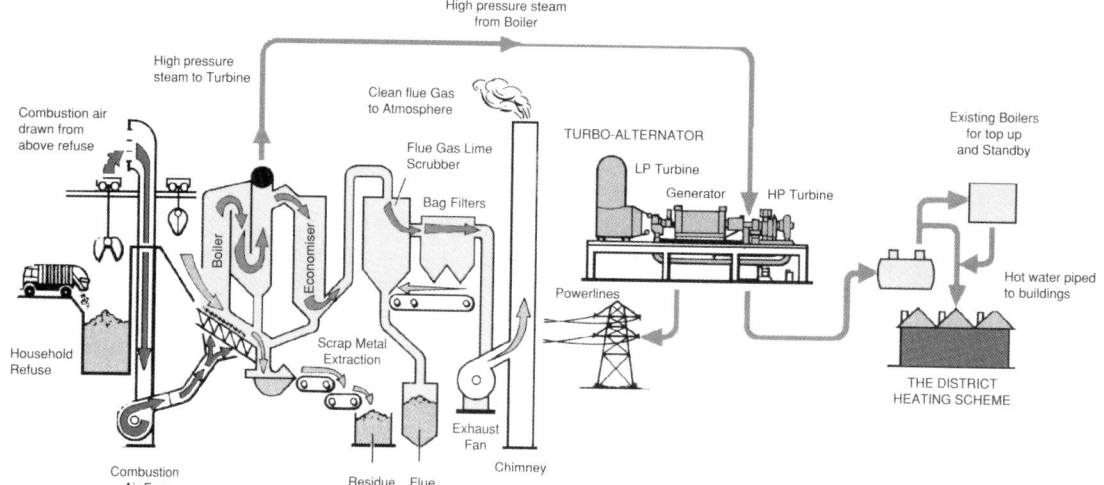

High pressure steam
from Boiler

High pressure
steam to Turbine

Combustion air
drawn from
above refuse

Clean flue Gas
to Atmosphere

Flue Gas Lime
Scrubber

Bag Filters

TURBO-ALTERNATOR

LP Turbine

Generator HP Turbine

Existing Boilers
for top up
and Standby

Boiler

Economiser

Household
Refuse

Scrap Metal
Extraction

Powerlines

Hot water piped
to buildings

THE DISTRICT
HEATING SCHEME

Combustion
Air Fan

Residue   Flue
Ash   Ash

Exhaust
Fan

Chimney

Figure 7.3
A schematic diagram of the SELCHP Mass-Burn Incineration Plant, Lewisham, in south-east London. (SELCHP)

plant in the United Kingdom is one such plant (Fig. 7.3). Here 420 000 tonnes of solid waste (with an approximate energy content of 1.8 GJ or 500 kWh per tonne) are collected each year, from 400 000 homes and from local businesses. This waste is fed into two incinerators, each capable of burning twenty-nine tonnes an hour. Steam produced by boilers is fed to a turbine driving a 31 MW alternator operating at 11 kV. Excess heat from the boilers is fed, as hot water, through six kilometres of doubly-insulated underground pipes, and used in a district heating scheme to heat 7500 homes and some local schools. Gases from the combustion pass through an advanced flue-gas cleaning plant to remove particulates, dust and acidic gases. The plant reduces the requirement for landfill space, recovers 20 000 tonnes of scrap metal for recycling each

year, creates fifty jobs and generates 30 MW of electricity, which is enough to supply 50 000 homes.

Biomass is at present the world's fourth largest energy source. It is a carbon dioxide-neutral fuel, in so far as the carbon dioxide liberated in the combustion of biomass fuel is only being returned to the atmosphere, at the most, a few years after its removal by the photosynthetic process. The greater use of biomass has the potential to reduce our dependence on fossil fuels, possibly eliminating their use for liquid transport fuel. New Zealand has great expertise in the growing of biomass. Conversion of some of our land to biomass production would not only increase employment opportunities, but would aid us in our obligations to stabilise carbon dioxide emissions.

8 Hydro Power

Humans have used the energy of flowing water since prehistoric times. During this century we have developed the technology to convert the energy available into electrical energy.

Solar energy, which is absorbed by the upper layers of the oceans, causes water to evaporate. Over New Zealand, prevailing westerly winds blow moist air onto the land. It is cooled as it rises over the mountains, and the water vapour it contains may turn to rain or to snow.

Water falling on high ground possesses gravitational potential energy. If, as it runs back towards the sea, it can be made to flow through turbines attached to electrical generators, then some of its potential energy can be converted through kinetic energy into electrical energy. We construct 'hydro-schemes', generating 'hydro power', for this purpose.

For a hydro-scheme to be effective, it needs a catchment area with adequate rainfall, a good head of water, and to be situated on a river which flows consistently throughout the year. Without the latter, it requires a storage lake with water sufficient to run the generators during periods of low rainfall, thereby maintaining continuity of supply. The Clyde dam is really a 'run of the river' dam. Lake Dunstan, which has been formed behind it, exists more to provide a head of water than for water storage. The level of Lake Dunstan varies by about 0.5 m each day, as water is used for generation. Lakes such as those at Pukaki and Tekapo are true storage lakes, and their levels can vary by several metres during the year, depending on seasonal rainfall and electricity demand.

New Zealand is lucky to have many such areas, and several fast-flowing rivers. At present, hydro-schemes generate about 78 PJ (21 TWh) of electrical energy each year, which is three-quarters of the electrical energy generated in this country. It has been estimated that this yearly total could be trebled to 230 PJ (64 TWh), if all the sites which have been identified as suitable for hydro-schemes were to be developed.

If we are to develop further hydro-schemes, these will have to meet not only the requirement of good rainfall and water storage, but also have low environmental and social impact. Other considerations are that they be either conveniently sited with respect to the demand for power, or sufficiently large to make the construction of long grid connections economic.

Most suitable hydro sites in the North Island have already been developed. Eight dams generating a total of 1040 MW are located on the Waikato river, and any further development of this river would destroy tourist attractions such as the Huka Falls. Many other rivers flow through national parks or, like the Wanganui river, have considerable cultural and recreational importance. It is probable that development of those North Island sites which would be socially acceptable and could generate electrical energy for less than 10c $(kWh)^{-1}$ would only contribute about 300 MW of new capacity.

By contrast, the South Island has many undeveloped sites, particularly in the Canterbury, Otago and West Coast regions. Here, over 2000 MW of capacity expected to be capable of generating energy for less than 10c $(kWh)^{-1}$ could be constructed. Even so, the estimated costs of these pro-

Figure 8.1
New Zealand's electricity generation sources, August 1995.
(ECNZ)

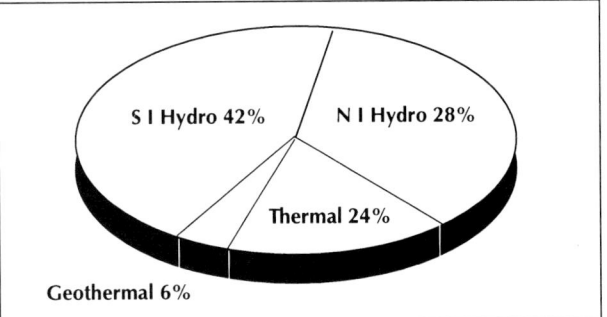

S I Hydro 42%   N I Hydro 28%

Thermal 24%

Geothermal 6%

posed hydro-schemes cover only engineering works and land purchase, and do not include other more intangible costs, which we will discuss later.

The history of the last major South Island hydro-scheme to be constructed, the Clyde Dam, is well documented. The construction of this dam involved some cost over-run. The possible final cost of $1.7 billion was numerically three times the original estimate, although when inflation is taken into account the dam itself was constructed for about its 1977 estimate. In addition to this sum, $337 million was spent stabilising the land adjacent to Lake Dunstan; this was not included in the original estimates. No other hydro-scheme undertaken in this country has met with problems of this scale. At the time of writing, a proposal by ECNZ to build a 350 MW hydro-scheme at Tuapeka Mouth, further down the Clutha, is meeting both local and national opposition. In fairness, it must be remembered that any major development will meet criticism, some of which is philosophic and some of which stems from self-interest. However, this proposal does have very serious environmental impact implications.

In terms of water flow, the Clutha is New Zealand's largest river, having an average flow rate close to 1000 cusecs at its mouth. The Clutha has a total generating potential of 1365 MW, and dams have already been constructed at Clyde (432 MW) and Roxburgh (320 MW), in both cases flooding land. If the Tuapeka scheme were to go ahead, a third lake would be formed, flooding the town of Beaumont, and farms and orchards nearby. The lake would also flood the highway to Central Otago, necessitating the construction of a roadway on higher ground. Furthermore, it would destroy Birch Island.

Unlike wind farming, hydro development usually destroys productive areas of land. When comparing alternative schemes, the ongoing loss of produce from this land should be included in the cost of electric-

Left: The town of Cromwell before the Clyde Power Project. (Photo: Randall Photography)

Above: Cromwell after the completion of the Clyde Dam. (Photo: John Fridd)

The Energy of the Clyde Dam

Construction of the Clyde Dam began in 1977, and was completed in 1994, at a cost of $1.7 billion. This hydro-dam converts the gravitational potential energy of water, through the rotational kinetic energy of a turbine, into electrical energy. The figure below shows cross-sections of the dam and spillway.

The dam has a final design capacity of 610 MW, but at present can generate 432 MW from four turbine-generator units. It was constructed of $1.2 \times 10^6 \, m^3$ of concrete, has a crest length of 490 m and has a height of 60 m measured from the original river level, its deepest foundation being more than 100 m below the crest. The construction of the dam formed Lake Dunstan, which has a surface area of 26.4 km^2 When generating at full capacity, 850 m^3 of water (850 tonne or 8.5×10^5 kg) flows through the turbines every second. This is said to be a flow of 850 cumecs. The lake's surface is held at a mean height of 54 m above the downstream river level, giving a design 'head' of 54m. The head is being increased to 58m by deepening the river downstream from the dam.

In order to see how this energy conversion occurs, we note that the gravitational potential energy of the water is calculated from the relationship

$$E_p = mgh$$

where m is the mass and h is the head of the water. In this manner if g is taken as 10 N kg^{-1}

$$E_p = 8.5 \times 10^5 \times 10 \times 54 \text{ J}$$

of energy is available for conversion to electrical energy every second, which is

$$459 \times 10^6 \text{ J}$$

Remembering that 1 W = 1 Js^{-1}, this implies that the Clyde Dam could generate electrical energy at 459 MW. However no dam has an efficiency of 100 per cent, as there are losses due to turbulence and in the generating system. ECNZ informs us that the Clyde Dam is over 90 per cent efficient overall, so if we assume that the efficiency of the whole process is 94 per cent, then the output power will be

$$459 \times 0.94 = 432 \text{ MW}$$

There is provision to install two more turbines in the generating house, which would increase the power of the dam to 610 MW, and make the Clyde Dam the most powerful hydro dam in New Zealand.

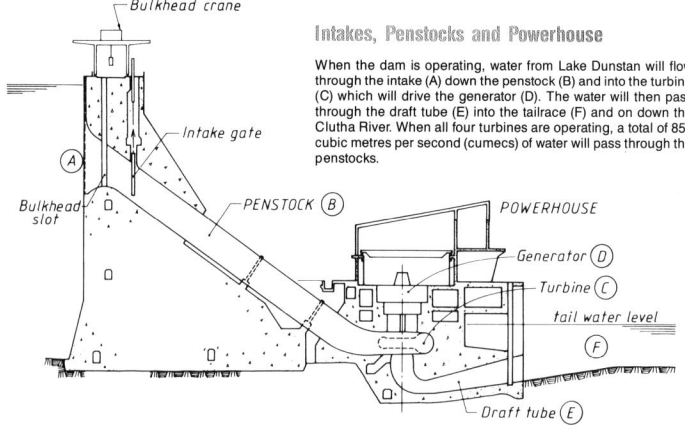

Intakes, Penstocks and Powerhouse

When the dam is operating, water from Lake Dunstan will flow through the intake (A) down the penstock (B) and into the turbine (C) which will drive the generator (D). The water will then pass through the draft tube (E) into the tailrace (F) and on down the Clutha River. When all four turbines are operating, a total of 850 cubic metres per second (cumecs) of water will pass through the penstocks.

(ECNZ)

ity from the dam, as without the dam it is likely that the flooded valley area would have been agriculturally more productive than the area around the new shoreline. Although the flooding of land does present economic and social disadvantages, if a large lake were created on the lower Clutha it might be able to support irrigation schemes, recreational activities or even be used for fish-farming. As an example of potential benefits, the town of Cromwell has developed a considerable tourist trade since the formation of Lake Dunstan, and section prices around the new shoreline are among the highest in the country. Cromwell, however, is situated in a particularly favourable climatic region; it is extremely unlikely that a development on the lower Clutha would provide long-term benefit for the local community in a way similar to what is occurring at Cromwell.

Construction of large hydro-schemes takes up to ten years, during which time a large workforce is employed. Unfortunately, unless another large scheme is then started, the workforce has to be disbanded. This has happened in the case of the Clyde Dam, and unless ECNZ or some other

Press report from the Otago Daily Times, *7 July 1995.*

Inefficient power stations 'cost' taxpayer $10 billion

Wellington (PA). — Taxpayers have lost more than $10 billion from bad decisions on economically inefficient power stations, according to a Treasury report.

The waste of public money is equivalent to the economy being poorer by between $430 million and $561 million a year, according to the McLaughlan report, made available on Wednesday by Electricity Supply Association analyst Alan Jenkins.

The continuing costs stem from taxpayers making up the difference between the true cost of power generation and what people pay.

The McLaughlan report was commissioned by Treasury in 1983 to calculate the true cost of power generation.

Figures from the report were converted to 1995 dollars by Mr Jenkins, using Ministry of Works construction cost indexes which include adjustments for inflation.

The report blames poor planning, construction delays and cost over-runs for additional expenses that were written off instead of recovered through power prices.

The report covers 11 of 17 power stations built between 1953 and 1983. It shows that none of the 11 stations is economically efficient, with true costs of power generation up to five times higher than what ECNZ is charging in today's dollars.

It shows the Atiamuri power station on the Waikato River has a true cost of power generation of 21.1c a unit. The Tongariro power scheme is even more expensive, with a true cost of about 25c a unit. Electricity from the Clyde dam costs between 12-17c a unit.

ECNZ charges an average wholesale price of 4.9c a unit around New Zealand.

The difference has been made up by taxpayer subsidies and accounting write-downs.

According to the report, the Tongariro schemes involved unnecessary extra costs of $1.3 billion. The Marsden-B and Whirinaki stations cost $1.87 billion too much while Clyde was about $1.3 billion too expensive. Clyde is now valued at $782.93 million.

In total, the money wasted has resulted in economic losses of $11.7 billion in today's dollars.

ECNZ spokesman Bruce Thompson said the corporation would not comment because the figures in the report challenged ECNZ's asset valuations. ECNZ bought its power stations, plus Trans Power, from the government for $6.3 billion in 1987.

Former ECNZ chairman John Fernyhough, who negotiated that deal, said on Wednesday that was a fair price because the asset values were based on the earnings potential of the power stations rather than their historical costs.

Mr Fernyhough acknowledged that previous governments had made huge blunders which had led to inefficient allocation of resources and some power stations should never have been built.

Electricity Supply Association executive director Barry Leay blamed politicians setting power prices for the financial disasters.

"They were plucking numbers out of the air," he said. This would change before the end of the decade when power prices and power station values would be set according to their true costs.

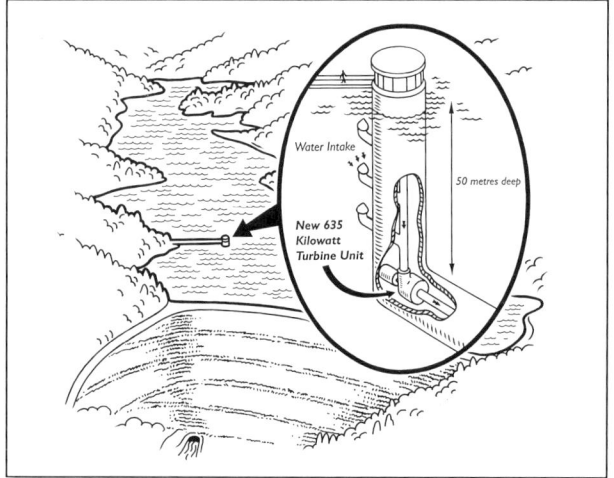

Small hydro-electricity generating scheme operated by Watercare Services on the Mangatangi storage lake near Auckland. The 635 kW plant produces enough electricity to supply a medium-sized township. (Watercare Services Ltd)

present.

Small schemes can be constructed to utilise energy which would otherwise go to waste. Auckland's water supply is now used to generate electricity. In November 1994, a 635 kW plant was installed in the water outlet of the Mangatangi storage lake in South Auckland. The energy generated is enough to supply about 250 homes.

Small hydro-schemes are also valuable in remote locations. The owners of the Tira Ora Tourist Lodge in the Marlborough Sounds have installed a 5.8 kW small hydro unit to supply electrical energy to the lodge, replacing a diesel generator which has been kept as a back-up unit. In its first year of operation, the generator produced 31 682 kWh at a cost of 14.5c (kWh)$^{-1}$, significantly less than the cost of the diesel system, which was 42.9c (kWh)$^{-1}$. At that rate the scheme, which had a capital cost of $40,000, will pay for itself in three years. Potentially small hydro-schemes could contribute as much as 500 MW, which represents the output of a major dam, to our generating capacity.

New Zealand has great experience in the construction and operation of hydro-schemes. The expansion of our existing hydro-capacity could contribute a further 150 PJ of electricity to our energy supply each year. This would be enough to cope with increasing energy demands for the foreseeable future. In operation, hydro-electric generation is clean, non-polluting and cheap. It has a good safety record, and the useful life of dams (which is more than fifty years) means that the capital cost of a dam is paid back several times within its working life. Against this, the development of large schemes such as that at Clyde leads to social disruption within the area of the scheme, and alters the environment considerably. We should not wish all of our scenic rivers to be developed for hydro-schemes, because many of our rivers are considered by most New Zealanders to be more valuable if left in their natural state,

contractor is able to provide a programme of hydro-development, the prospects for consistent employment are poor.

Although we usually think of hydro-schemes in terms of massive dams such as those at Clyde or Benmore, there are a number of hydro-schemes in New Zealand which have outputs in the range of 10 to 100 MW and which serve local communities or are attached to the national grid. As well, there are a number of small schemes, often producing only a few kW, which either serve individual consumers or are operated by electricity marketing utilities. Although small hydro-schemes might not seem important, they contribute 126 MW to our country's economy at rather than altered for the purposes of electrical generation. On the other hand, many of our present hydro-lakes provide tourist and recreational facilities, as well as stored water for irrigation schemes: this has created employment within the communities close to the scheme.

Probably the greatest drawback of the further development of hydro is that the best sites are in the South Island, far from the demand, and the energy produced would have to be transferred across Cook Strait. This would entail capital cost in upgrading the national grid and the Cook Strait cable. As well as this, the distance from the South Island to the main population centres of the North Island would add 20 per cent to the delivered cost of the energy. In view of these constraints, and unless there were a major increase in the South Island's demand for energy, it may well be that cost analysis, taking into account social and environmental factors, will show that further major hydro-development in the South Island is at present not appropriate, and that we should pursue other alternatives.

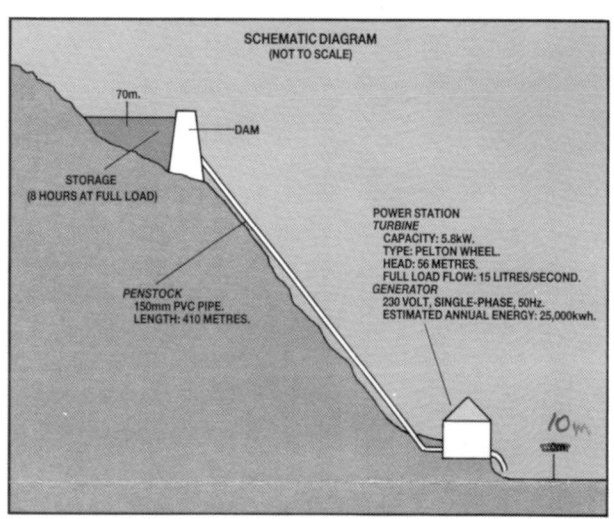

SCHEMATIC DIAGRAM
(NOT TO SCALE)

70m.

DAM

STORAGE
(8 HOURS AT FULL LOAD)

POWER STATION
TURBINE
CAPACITY: 5.8kW.
TYPE: PELTON WHEEL.
HEAD: 56 METRES.
FULL LOAD FLOW: 15 LITRES/SECOND.
GENERATOR
230 VOLT, SINGLE-PHASE, 50Hz.
ESTIMATED ANNUAL ENERGY: 25,000kwh.

PENSTOCK
150mm PVC PIPE.
LENGTH: 410 METRES.

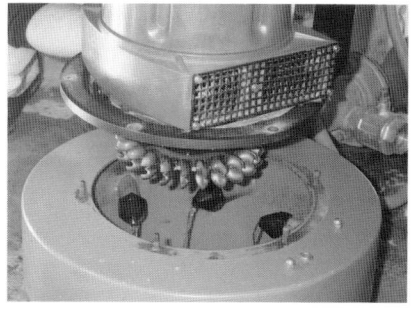

The Tira Ora turbine and generator removed from their scroll case (visible below the generator is the pelton runner and three pelton nozzles). (Photos and diagram: Marlborough Electric)

A diagram and photographs of the Tira Ora Lodge micro-hydro scheme in the Marlborough Sounds. Above: storage pond and airstrip. Left: powerhouse operating at full load (5.8 kW).

9 Geothermal Energy

The current theory of the formation of the Solar System suggests that it was formed 4.5 billion years ago when a supernova explosion compressed interstellar gas, causing it to begin to collapse. The intense temperature ($>10^8$ K) at the core of the supernova fused the nuclei of light elements such as carbon and silicon, creating elements with atomic numbers up to the region of 90. Some of the debris of heavy elements ejected from the supernova mixed with the collapsing gas cloud. The planets were formed by the accretion of small masses of this debris, known as planetesimals. As these planetesimals impacted on the growing Earth, kinetic energy was converted to heat energy, and the young Earth became molten. Dense metals, particularly iron and nickel, settled to the core of our planet.

General view of Wairakei geothermal station. (Photo: ECNZ)

If no further heat energy had been added to the interior of the Earth it would have cooled within 3×10^8 y. In fact, radioactive materials such as thorium, actinium and uranium, with half lives in the order of 10^{10} y, are present in the Earth's core, and their decay releases prodigious quantities of heat, estimated to be more than 3.6×10^{23} J (0.36 YJ or 100 PWh) each year. The geothermal source represents an energy supply 10^5 times greater than the world's primary energy demand.

Although more than half of the original radioactive elements will have decayed in the 4.5×10^9 y since our planet formed, radioactive decay in the Earth's core and mantle shell keeps the core at high temperature and ensures that rocks close to the surface are still warm. As this warming is constant over human time-spans,

geothermal energy, provided it is not depleted by local over-use, may be regarded as renewable.

On average, the temperature of the Earth's interior increases by 3° C for every 100 m depth from the surface. This means that at a depth of 3 km the temperature of the earth is about 100° C. If geothermal heat were conducted from the Earth's core evenly over our planet, it would be too diffuse a source to be useful. Fortunately, in certain locations, notably at tectonic plate boundaries, hot magma rock lies close to the Earth's surface or during volcanic activity actually emerges. If the hot rock is close to underground water supplies (aquifiers), this water can be superheated under pressure to temperatures in excess of 200° C. New Zealand is one such place where this occurs. The Rotorua-Taupo region, which is about 250 km long and 50 km wide, is one of our planet's most geologically active places.

Geothermal energy was first tapped for electrical energy production in 1913, when a geothermal power station was constructed at Lardarello in Italy. Wairakei, near Taupo, was the second geothermal plant to be constructed and was commissioned in 1963. At the Wairakei geothermal station, sixty-one wells of an average depth of 610 m and a maximum depth of 1500 m reach down to a vast underground water system which has been heated, at great pressure, to temperatures in excess of 220° C. Steam is produced when the bores release the pressure, allowing the water to boil. The steam is brought to the surface, any water contained in it is separated and the resulting 'dry' steam drives the power station's turbines. These are housed in two powerhouses and have output capacities of 102.6 and 90 MW respectively. A schematic diagram of a dry steam system such as the Wairakei plant is shown in Fig. 9.1.

The Wairakei power station produces 4 PJ (1 100 GWh) of electrical energy annually, which is about 5 per cent of our

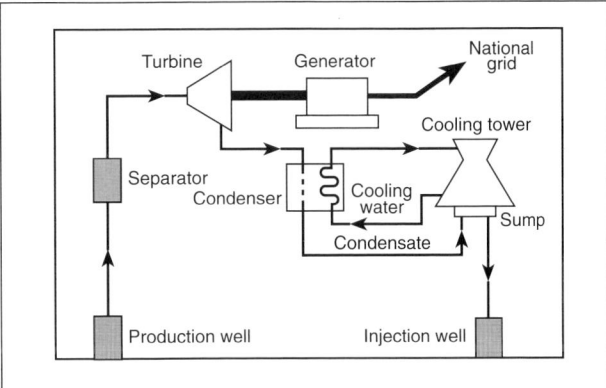

Figure 9.1
Generation of
electricity using
the dry steam
system employed
at Wairakei.

total electrical energy production or 10 per cent of the North Island's demand. The Wairakei field has a high capacity factor of around 85 per cent, which means that geothermal energy is very suitable for base-load energy production and also that the energy is produced cheaply.

New Zealand's geothermal resource is large. Although the Wairakei field is fully developed, a further 1100 MW of capacity, with a probable annual output of 10 000 GWh (36 PJ), has been identified in the Taupo-Rotorua and the Ngawha (Northland) volcanic regions. This resource would supply one third of our present electrical energy demand if it could be fully developed.

Although geothermal energy has a high capital cost of $4000 (kW)$^{-1}$, its high capacity factor, long plant life and zero cost of fuel allows electrical energy to be generated at 6c (kWh)$^{-1}$, which is a very attractive price in today's terms.

From the point of view of the environment, geothermal fields do have some disadvantages. The disposal of waste water and heat can cause problems, as direct disposal into local streams is unwise. However, if use is made of the waste hot water in some enterprise such as fish-farming, not only is this problem avoided, but the overall cost of geothermal energy is lowered, possibly to 4c (kWh)$^{-1}$. Another disadvantage is local noise and steam emis-

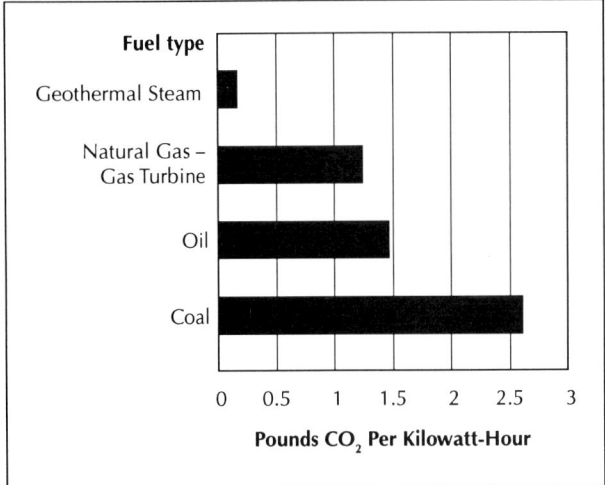

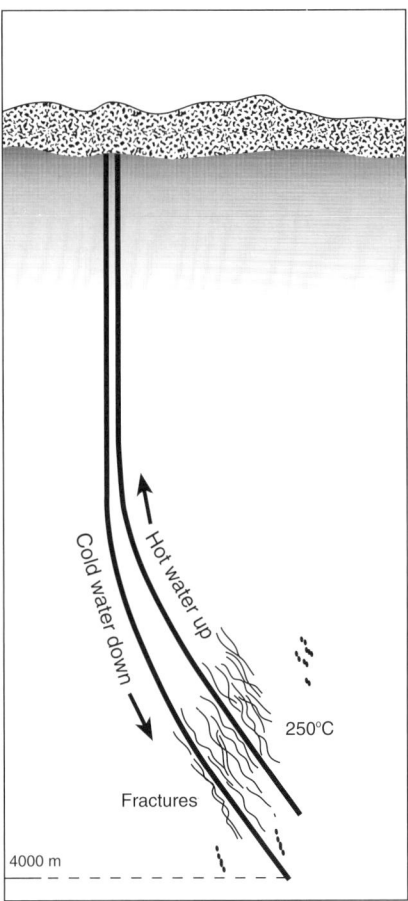

Figure 9.2
Relative CO$_2$
emissions of geo-
thermal and fossil-
fuel electrical
generation.
(World Energy
Council)

sion, the latter leading to fog formation. As well, geothermal drilling releases carbon dioxide and sulphur dioxide from the hot rock strata, which diminishes the greenhouse benefits and produces acid rain. A comparison of the CO$_2$ emissions of geo-thermal and fossil-fuel electrical generation is shown in Fig. 9.2.

As the energy is available close to the large cities of the North Island, particularly Auckland, the energy losses due to transmission are considerably reduced, providing an advantage over South Island hydro generation. As most geothermal areas are significant to Maori, then any development of the resource would have to be made with due regard to the wishes of the tangata whenua.

The expertise of New Zealand engineers in geothermal development has earned worldwide respect. This expertise is 'exportable', especially to developing regions such as Indonesia, where there is a high geothermal potential. This presents employment and foreign exchange advantages for our country, which although not directly quantifiable in economic terms are positives for further geothermal development.

Further into the future it may be possible to tap the energy of hot dry rock or even magma, by drilling down to rock

Figure 9.3
An idea for the future – tapping the energy of hot dry rock by injecting water through a drill-hole.

which has temperatures well in excess of 100° C, fracturing this rock and injecting water down the drill-hole. Steam would then be extracted through a second bore. This technique, which is illustrated in Fig. 9.3, would extend our reserves significantly.

Research into hot dry rock energy began in the US in 1972 and a pilot plant has been operating since 1992. A complementary research programme in Cornwall (UK) has recently been suspended due to lack of progress, but projects in the Ukraine, Europe and Japan have also been initiated.

10 Nuclear Energy

Due to the deep-seated anti-nuclear beliefs of the New Zealand population, the use of nuclear energy is no longer a political option for New Zealand. However, it is an available source of energy which could contribute to a sustainable energy policy, and so it is considered here, for information purposes only.

Although not a renewable energy form, nuclear energy does not utilise fossil fuels and, excepting contributions due to the mining and smelting of uranium, does not contribute carbon dioxide nor indeed any greenhouse gas to the atmosphere during its operating cycle.

Nuclear power plants were developed during the years immediately following World War II. It was hoped at the time that they would provide 'electricity too cheap to meter'. Unfortunately, to date, this has not been realised. Nevertheless to 1993, seventeen countries have constructed non-military nuclear power plants. This represents a capacity of 90 GW from 430 plants and is 15 per cent of the global electrical generation capacity. In the United States, nuclear energy accounts for 20 per cent of electricity generation and in France this figure is 78 per cent. In Germany, it is said that 2 billion tonnes of CO_2 emission have been saved since the introduction of nuclear energy in 1961. What, then, are the arguments for and against nuclear power?

Nuclear power stations do not contribute significantly to the production of greenhouse gases. This is clearly useful, for if global warming takes the course predicted, loss of life might be measured in millions, and damage in billions, of dollars. Neither do they emit SO_2 or NO_x. However, a substantial amount of low-grade waste heat is generated and this is often discharged as warm water into rivers or, as at San Clemente, into the sea, modifying the ecosystem.

Although its output cannot be altered

Nuclear power station at San Clemente, California (USA).

Nuclear Reactions and Reactors

The nuclei of elements with atomic numbers greater than eight-three are radioactive and decay to lighter elements by a variety of radioactive decay routes. The cause of this nuclear instability is the mutual repulsion of the large number of protons inside the nucleus. In the case of certain nuclei, the decay may occur by the splitting of the nucleus into two approximately equal halves. This process is called nuclear fission and substances are called fissile.

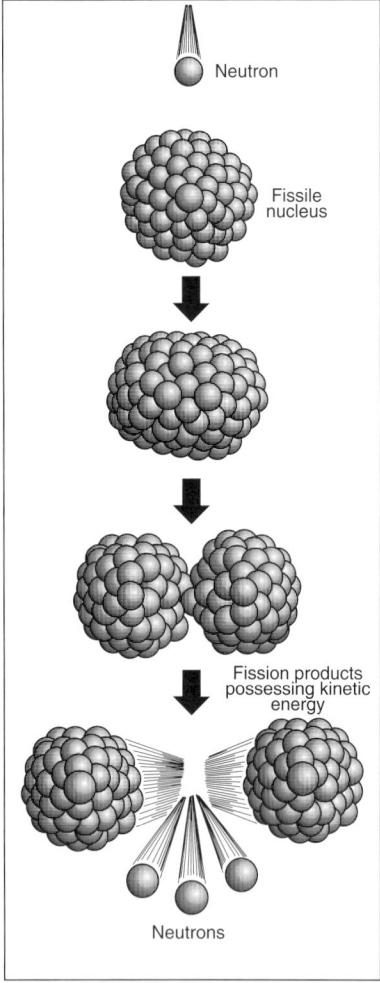

Figure 10.1
A schematic view of the neutron-induced fission of uranium-235.

Uranium-235 ($^{235}_{92}U$) and plutonium ($^{239}_{94}Pu$) are two common fissile substances. A fission reaction is represented schematically in Fig. 10.1.

When fission occurs, the smaller nuclei formed have lower proton-proton repulsion forces compared to the original nucleus, and some of the energy which held that larger nucleus together by the strong nuclear force is no longer needed and may be released to the surroundings.

Natural uranium contains 0.7 per cent of uranium-235 which can be concentrated, usually to about 4 per cent to give 'enriched uranium'. The remainder is uranium-238 ($^{238}_{92}U$) which decays by release of an alpha particle ($^{4}_{2}He$),

$$^{238}_{92}U \Rightarrow ^{234}_{90}Th + ^{4}_{2}He$$

In a nuclear reactor, fission reactions occur within the fuel rods. A neutron formed in a previous fission reaction may strike a uranium-235 nucleus, causing it to undergo fission. One typical reaction is indicated by the equation below:

$$^{1}_{0}n + ^{235}_{92}U \Rightarrow ^{90}_{38}Sr + ^{142}_{54}Xe + 3^{1}_{0}n$$

but it must be understood that a wide variety of product nuclei are formed and this is only one of the possible reactions occurring.

These disintegrations release more neutrons than are absorbed. The neutrons travel away at high speed, and would probably leave the reactor core or else become absorbed by other nuclei in the core. If this were to happen a chain reaction could not be maintained.

The fuel rods are surrounded by a moderator, usually water or graphite, which slows the neutrons down. Slow neutrons are more likely to be absorbed by uranium-235, initiating a fission event.

If for each fission one of the ejected neutrons is captured by another uranium-235 nucleus, then the reaction becomes self-sustaining and may be maintained for long periods of time. If too many neutrons are present in the core, then control rods are inserted to reduce their number. These control rods usually contain boron, which absorbs neutrons,

$$^{10}_{5}B + ^{1}_{0}n \Rightarrow ^{4}_{2}He + ^{7}_{3}Li$$

Most of the neutrons which strike the uranium-238 are absorbed, resulting in the formation of plutonium-239,

$$^{238}_{92}U + ^{1}_{0}n \Rightarrow ^{239}_{92}U \Rightarrow ^{239}_{93}Np + ^{0}_{-1}e$$

$$^{239}_{93}Np \Rightarrow ^{239}_{94}Pu + ^{0}_{-1}e$$

This plutonium can be extracted from spent fuel rods and is itself a valuable fissile material.

Fig 10.2 displays the layout of a typical nuclear power station. Heat is removed from the reactor core by either a liquid or gas and is used to generate steam, which in turn rotates a turbine connected to the electrical generator. A typical nuclear power site will have several reactors.

The energy obtained from the fission reaction can be

calculated using the Einstein mass-energy equivalance equation $E = mc^2$.

Where c is the speed of light 3×10^8 ms^{-1}. The calculation is quite simple. The initial masses of the neutron and the fissile nucleus are added and the masses of the fission products and ejected neutrons are subtracted from this to give the mass defect (kg). This mass defect when multiplied by the square of the velocity of light gives the liberated energy (J).

An example of this type of calculation is provided by the 1993 University Bursaries Examination Question 13:

Plutonium can undergo a fission reaction when a neutron collides with a nucleus. The products of a typical fission reaction are strontium, barium and neutrons as shown by the equation:

$$^{239}_{94}Pu + {}^{1}_{0}n \rightarrow {}^{93}_{38}Sr + {}^{142}_{56}Ba + 5{}^{1}_{0}n$$

Fission reactions are used to produce energy for electric power generation.

(d) Use the information below to calculate the energy released in this reaction.
The velocity of light, c = 3.00×10^8 m s^{-1}.

Rest masses: (4 marks)

$^{239}_{94}Pu$	396.929 35 $\times 10^{-27}$ kg
$^{93}_{38}Sr$	154.278 37 $\times 10^{-27}$ kg
$^{142}_{56}Ba$	235.642 16 $\times 10^{-27}$ kg
$^{1}_{0}n$	1.674 83 $\times 10^{-27}$ kg

(e) Benmore hydro-electric power station in the South Island of New Zealand, has a power output of 500 MW. One tonne of plutonium contains 2.53×10^{27} atoms. If all the plutonium was made available in the fission reaction, calculate the time in years for Benmore to generate the equivalent energy.

(2 marks)

In order to solve this question, we first calculate the total mass of the plutonium-239 nucleus and neutron

$3.9846218 \times 10^{-25}$ kg.

From this, we subtract the combined masses of the nuclei of strontium-93 and barium-142 and five neutrons which is $3.9829468 \times 10^{-25}$ kg,

to give the mass defect 1.675×10^{-28} kg.

Multiplication of this mass defect by c^2 (= 9.00 x 10^{16} m^{2}s^{-2}), gives a value for the energy released for one fission reaction, which is

2.8×10^{-11} J

Multiplication of this energy by 2.52×10^{27}, the number of plutonium atoms in one tonne of plutonium, gives the energy released from the complete fission of one tonne of plutonium which is 7.1 x 10^{16} J (71 EJ).

Now Benmore dam generates 500 x 10^6 J s^{-1} (500 MW) and so would require

$$\frac{7.1 \times 10^{16}}{5 \times 10^6} \text{ s}$$

to generate 7.1 x 10^{16} J. This is 4.5 y.   (NZQA)

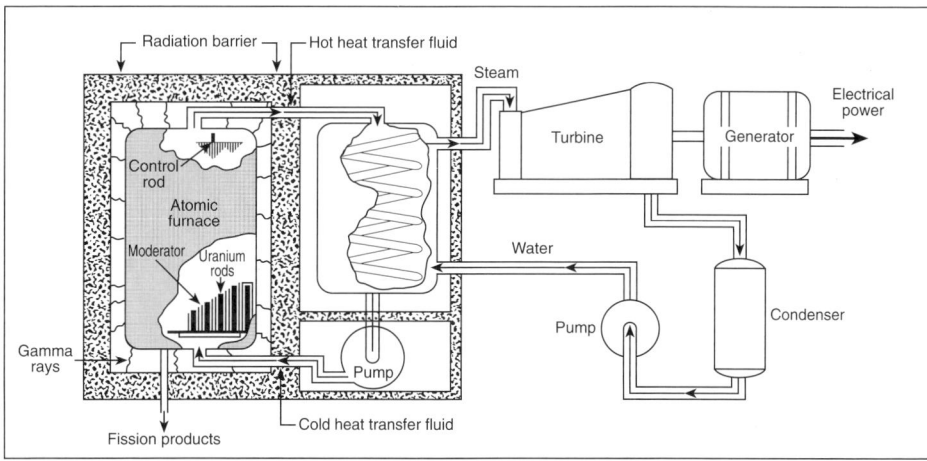

Figure 10.2 Layout of a typical nuclear power station.

Nuclear Fusion

Our Sun produces energy by the process of nuclear fusion, whereby two small nuclei unite to form a larger nucleus (see 'The Solar Resource', p.59). If a fusion reaction could be controlled, it would represent a source of energy.

Unfortunately the proton-proton reaction which occurs within the core of the Sun is unlikely to be of practical use on our planet. However, a reaction involving hydrogen-2 (deuterium : ^2_1H) and hydrogen -3 (tritium: ^3_1H) appears promising:

$$^2_1H + ^2_1H \Rightarrow ^3_2He + ^1_0n \qquad 5.32 \times 10^{-13} J$$

$$^2_1H + ^2_1H \Rightarrow ^3_2He + ^1_0n \qquad 5.32 \times 10^{-13} J$$

$$^2_1H + ^3_1H \Rightarrow ^4_2He + ^1_0n \qquad 28.14 \times 10^{-13} J$$

These reactions liberate neutrons which can be stopped by concrete barriers. Neutrons decay rapidly, the decay reaction having a half-life of 1000s.

$$^1_0n \Rightarrow ^1_1p + ^0_{-1}e + v$$

Neutron decay forms no other radioactive materials, and so nuclear fusion presents no environmental risk. However, neutrons can be absorbed by the nuclei of atoms forming radioactive isotopes. It has been suggested that even if fusion reactors can be successfully constructed, because of the intense production of neutrons by the fusion reaction, and their reaction with the nuclei of the atoms comprising the concrete shield, the concrete of the buildings which house the reactor will eventually become radioactive and therefore a health hazard. This would not only limit the useful life of the reactor, but also present problems with the disposal of the radioactive building material.

The energy liberated by the reaction could easily be absorbed in molten lithium, which would act as a heat exchange fluid and also absorb neutrons resulting in the generation of hydrogen -3:

$$^1_0n + ^7_3Li \Rightarrow ^3_1H + ^4_2He + ^1_0n(slow)$$

$$^1_0n(slow) + ^6_3Li \Rightarrow ^3_1H + ^4_2He + 7.7 \times 10^{-13} J$$

The hot molten lithium would then flow to a heat exchanger, transferring its heat to water, raising steam, which in turn would drive a turbine to generate electrical energy.

Deuterium makes up 0.02 per cent of all hydrogen on our planet and so 2 g of deuterium could be extracted from each 100 litres of sea water at an estimated cost of $10. In this manner, the fuel for such a reactor would be practically free and inexhaustible.

As the reactions only occur at temperatures in excess of 10^8 K, it might seem that such temperatures are impossible to obtain on our planet. However, two approaches seem promising. One approach, the 'Tokamak', confines the nuclear fuel in a doughnut-shaped apparatus by a magnetic field, and then heats the material by some means such as radio frequency radiation. Experimental devices of this type have recently produced more energy than was needed to heat the plasma initially, a major breakthrough. An alternative approach is to focus several lasers simultaneously on a pellet of fuel containing hydrogen-2 and -3. This has the effect of causing the surface of the pellet to evaporate, producing a shock wave which not only compresses the core of the pellet but also heats it to the required temperature, igniting the fusion reaction.

At present the engineering feasibility of these fusion reactions is not proven. However, a great deal of progress has been made. In terms of engineering, the development of nuclear fusion reactions will be the greatest undertaking yet attempted by the human race. If a successful fusion reaction can be developed during the twenty-first century, it is reasonable to hope that unlimited energy could be produced. Nevertheless the capital cost of this energy could be such that fusion reaction might only be constructed in the major industrialised countries, and that smaller countries such as our own will still rely on other energy sources.

quickly as load changes, a nuclear power station of 1000 MW capacity could provide the base electrical demand for the North Island, with hydro being used for 'peak clipping', allowing all fossil-fuel stations to be mothballed.

Australia has major reserves of uranium. Provided that we, or the Australians, developed the technology to extract and manufacture the fuel rods, it is probable that our supplies of nuclear fuel would then be unaffected by world political change.

All forms of power generation involve risks. The mining of coal may be accompanied by loss of life and damage to health, and causes environmental damage. Dams may collapse; one in Gajvrat in India did so, killing 1500 people. Nuclear power, by contrast, has a good safety record to date. The only major accident occurred in a reactor at Chernobyl; its design was always suspect, and would not have been licensed in the West. The Ukrainian Health Minister, Andrei Serdyuk, announced in 1995 that the death toll from this accident has risen to 125 000, and that rescue workers, children and pregnant women are still prone to radiation-induced diseases. More people than this die on the roads worldwide each year, but we do not ban the motor car.

It must be realised that we are always exposed to radiation. It exists as a background, due to radioactive materials in rock and soil, and is produced by cosmic radiation bombarding our atmosphere. We are particularly exposed to these during air travel. Our bodies contain potassium-40 and carbon-14, of which both isotopes are radioactive. The uranium that we mine is already present in the Earth's crust, and so we are merely concentrating what already exists. Furthermore, nuclear reactors are well shielded so that no radiation is likely to affect the general population. Modern reactors are very safe, having incorporated the lessons of accidents such as Chernobyl and Three Mile Island.

What, then, is the case against nuclear power? As has been indicated, the hope of cheap energy from nuclear reactors has not been realised. During a major public inquiry into a proposed nuclear station at Hinckley Point in the west of England, it became apparent that Lord Marshall, a major advocate of nuclear power, had been lying about the real cost of nuclear power. He subsequently had to resign as Chairman of the Central Electricity Generation Board (CEGB). When the real costs became public knowledge, nuclear stations were withdrawn from the privatisation of the British Electrical Generating industry. British banks let it be known that they would not invest in nuclear power, particularly because the costs of radioactive waste disposal and eventual decommissioning of plants appear likely to be equal to the cost of building and running plants during their lifetime. These banks have subsequently financed several British wind farms.

The British government introduced a Non-Fossil Fuels Obligation (NFFO) in 1989. This is a levy on fossil-fuelled energy production, which is used to subsidise production from non-fossil fuel sources. Ninety-five per cent of the NFFO money goes towards subsidising nuclear power plants. At the Hinckley Point inquiry, the CEGB admitted that wind power was considerably cheaper than nuclear power, a situation which has become more true today.

Although the risk of a nuclear accident is small, should such an accident occur the impact on New Zealand would be devastating. Disregarding loss of human life and increased likelihood of cancers if radioactive material contaminated a portion of the country, the loss to our farming industry would be crippling. We have a 'clean, green' image which is valuable to our farming exporters. Furthermore, we take great care to avoid the introduction of any organism or disease which might affect our farming industry. An accident even of a minor kind might effectively close down our multi-billion dollar meat exporting industry for many years.

It must be noted here that nuclear power stations worldwide cannot buy insurance against accidents; the private insurance industry considers them too risky.

A major part of the cost cycle of nuclear power stations is the problem of waste disposal and decommissioning. A nuclear reactor generates several cubic metres of radioactive waste every year. This must be stored until it ceases to be hazardous. Although the activity of the spent fuel drops rapidly, with only a few per cent of its initial activity remaining after one year, it is still necessary to store the waste for several hundred years before it may be considered safe.

At present, the best method for storage appears to be to seal highly radioactive waste into glassy beads or similar containers and store this waste deep underground, in regions of geologically stable rock which are free of seismic activity, an unlikely scenario for New Zealand. Low-level radioactive waste in both liquid and solid form is more difficult to deal with; unfortunately it represents a large proportion of the unwanted material. Sweden stores its waste in tunnels under the Baltic Sea, which is not a particularly attractive solution. If we were to construct nuclear power stations, we may find it necessary to ship our waste overseas for disposal, adding considerably to the cost cycle.

Finally dismantling the reactor at the end of its life presents problems. The United Kingdom is decommissioning the Windscale Advanced Gas Cooled Reactor. A three-stage decommissioning strategy is proposed. When the plant is shut down, the fuel and coolant are immediately removed, which reduces radioactivity at the site to 0.1 per cent of the initial level, although the fuel and coolant still have to be stored. Subsequently, most of the site is demolished over a 5-10 year timespan. The reactor vessel itself will then be left for 130 years before demolition. One suggestion here is to cover the vessel with a hill of soil. It might be pleasant then to plant a tree on this hill, giving the world a second One Tree Hill!

The Half-life of a Radioisotope

The nuclei of atoms are composed of two types of particles: protons and neutrons. Together these particles are referred to as nucleons. The protons within a nucleus repel each other as they are positively charged, however all nucleons 'feel' the strong nuclear force and so mutually attract. The stability of a nucleus depends on the balance between the repulsive force, which is electromagnetic in origin, and the strong nuclear force.

The number of protons within a nucleus determine the chemical nature of that element. Thus it is that all atoms with 92 protons within the nucleus are atoms of uranium, whereas all atoms with 90 protons in the nucleus are atoms of thorium. The number of nuclear protons is the atomic number of that element.

The total number of nucleons (protons and neutrons) within the nucleus is termed the mass number of that nucleus. These numbers may be included in the symbol for any particular nucleus thus:

$$\substack{\text{mass number} \\ \text{atomic number}}\text{element}$$

In this manner an atom of uranium might be written:

$^{238}_{92}$U or uranium-238

implying that in the nucleus, there exist 238 nucleons of which 92 are protons, and by subtraction, 146 must be neutrons.

Atoms of uranium exist which have differing numbers of neutrons within their nuclei. As they are atoms of uranium they still must possess 92 protons, but, for example, nuclei exist possessing 92 protons and 143 neutrons. These nuclear species have a mass number of 235 and may be written:

$^{235}_{92}$U or uranium-235

Atomic species of the same element which have differing numbers of neutrons and therefore mass numbers are termed isotopes.

Many isotopes are unstable and change to other nuclear species by a variety of decay modes, eventually arriving at species which are stable. The decay of an isotope obeys the radioactive decay law. This law states that when an isotope decays, it does so at a rate whereby the amount of isotope, and therefore its individual radioactivity, declines to half its original value in a time characteristic of that isotope. This time is called the half-life of the isotope.

Half-lives vary widely. Uranium-238 has a half-life of 4.5×10^9 y, whereas polonium-212 has a half-life of 3×10^{-7} s. A typical decay curve is shown in Fig. 10.3.

As can be seen from the graph, the activity, and so number of atoms, remaining after one half-life is one-

Figure 10.3 A typical decay curve.

half of the original. After two half-lives, one-quarter of the activity remains, and after three half-lives one-eighth of the activity remains. The process continues as long as any of the isotope has not decayed, but for practical purposes, related to the disposal of wastes, it is accepted that the material has decayed sufficiently to be safe for disposal after five half-lives, when its activity has reduced to $\frac{1}{32}$ or 3 per cent of the original value.

There are some points to note. Generally, the danger inherent in samples of very long-lived isotopes is less than that due to those with shorter half-lives, as the latter are decaying faster, and so emitting more radiation per unit of time. A second factor in dealing with isotopes produced in nuclear reactors is that many, such as isotopes of strontium, iodine or caesium, are able to be ingested by animals or humans and therefore represent a greater danger than isotopes which cannot become parts of our bodies. Finally, the danger of radiation depends on the type and energy of the radiation. Uranium-238 has a very long half-life, and does not become incorporated into our bodies. Furthermore, the alpha rays emitted by uranium-238 are relatively harmless, as they are generally stopped by the dead outer layers of our skin. On the other hand strontium-90 is a dangerous material, not only because it has a relatively short half-life of 28.1 y, but also because it behaves chemically like calcium, and can be absorbed by our bodies, particularly our bones, becoming an internal source of radiation. Finally, the beta radiation which it emits is not so easily stopped as alpha radiation and so can better penetrate into our tissue, damaging or destroying our cells.

11 Tidal, Wave and Ocean Thermal Energy

The Earth's oceans represent a vast store of energy in the form of the tides, waves and salt and temperature gradients. Indeed, the total extractable power represented by these energy sources might be in the order of tens of terawatts. New Zealand as an island country should consider the possibilities of harnessing this energy.

Tidal energy is caused by gravitational interactions between the oceans and the Moon, and to a lesser degree the Sun (Fig. 11.1). The Moon causes bulges with an amplitude of less than a metre on the near and far side of the Earth's oceans. These waves sweep round the earth in a westward direction with a period of 12.4 h. The Sun's contribution is to raise waves of smaller amplitude and a period of 12 h.

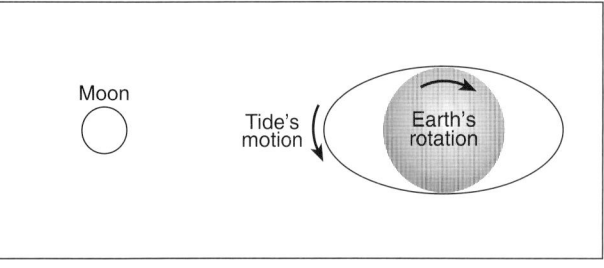

Figure 11.1
Tidal energy is caused by gravitational interactions.

When the Sun, Moon and Earth are in a line these waves add to give spring tides, whereas at half moon, when the Moon and Sun are at right angles as seen from Earth, the tides have a smaller range and are called neap tides. At certain places on the Earth's surface the tidal waves are concentrated to give large tidal ranges of up to 12 m, however around New Zealand the range is usually less than 2 m.

If we were to tap these tides, we would be extracting gravitational potential energy from the Earth-Moon system. However, the effect on this system would be insignificant. Indeed, energy is always being lost to friction, which causes the Earth's rotation to slow by 1 s every 100 000 y and the Moon to recede from the Earth.

Tidal energy has been used on the north-western European coast for one thousand years to drive mills. The technique involved the creation of ponds to trap water at high tide, which then drove a water wheel as the tide dropped. In 1966 the French constructed a barrage at La Rance on the north-west coast of France, where the tidal range is 8 m. The turbines, driven by the falling tide, have an installed capacity of 240 MW and a yearly output of 540 GWh (2 PJ). Since 1980, other experimental plants have been constructed in China, Canada and the former USSR, and proposals have included an 8.5 GW plant on the River Severn in western England.

Because of its small tidal range, tidal energy does not seem to have application to New Zealand at present. However it has been suggested that a giant underwater turbine might be moored at French Pass in the Marlborough Sounds, to extract tidal energy in much the same way that a wind turbine extracts energy from the wind.

Of greater significance for New Zealand is the energy available from ocean waves. These are caused by interaction between the wind and the ocean surface and so are derived ultimately from solar energy.

The west coast of the South Island is particularly exposed to large ocean waves which arrive continuously throughout the year, and so represent a constant and reliable source of energy, which might represent a yearly yield of 500 MWh (= 1.8 GJ) for each metre of shoreline.

There are several ways in which wave energy could be extracted. Prototype plants which may be moored off-shore and extract energy from their up-and-down motion have been tested. The 'Salter Duck' was an early device of this type. Alternatively, on-shore plants which direct the waves up tapered channels into reservoirs and then allow the water to run back to sea through a turbine have been con-

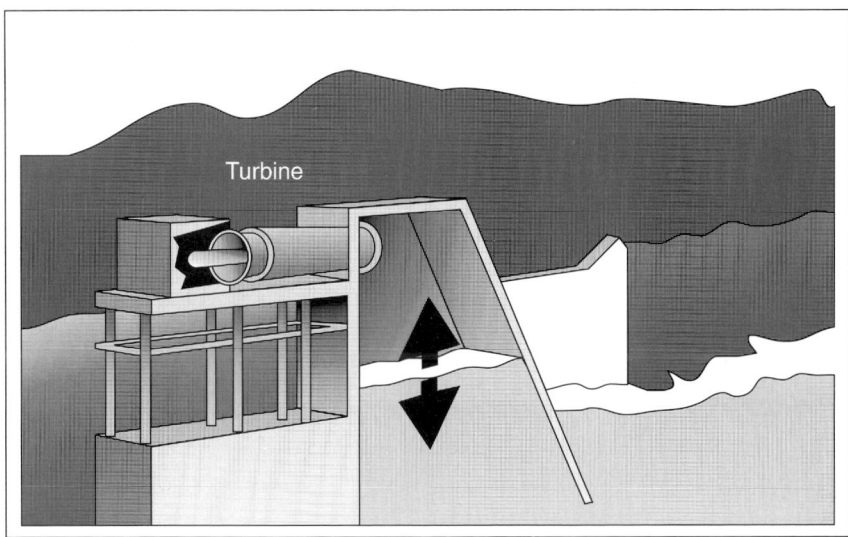

Turbine

Figure 11.2
Generator built to
extract wave
energy, on the
Scottish island of
Islay.

Figure 11.3
Prototype
commercial plant
designed for the
River Clyde at
Dounreay in
Scotland.

structed in Norway. A second type of shore-based machine allows water to oscillate in a chamber, the trapped air driving a turbine which has been designed so that it always turns the same way, whichever way the air is moving. A 75 kW generator of this type (Fig. 11.2) has been constructed on the Scottish island of Islay. It is connected to the grid and has been in operation for several years.

A prototype commercial plant has been constructed in northern Scotland and is to be positioned close to the Dounreay nuclear reactor (Fig. 11.3). This plant, which is twenty metres high, is designed to sit 350 m offshore in an average of 14 m of water. Two sand-filled ballast tanks, which anchor the device to the sea bed, funnel the waves into a collector tank where air is forced through one-way turbines, as in the Islay plant. It is expected that the plant will develop a maximum of 2 MW of electrical power, which will be supplemented by a 1.5 MW wind turbine attached to the structure. The generated cost of electricity is expected to be about 10c $(kWh)^{-1}$. Unfortunately, production of power by this prototype has been delayed, as it sank while being towed to its position.

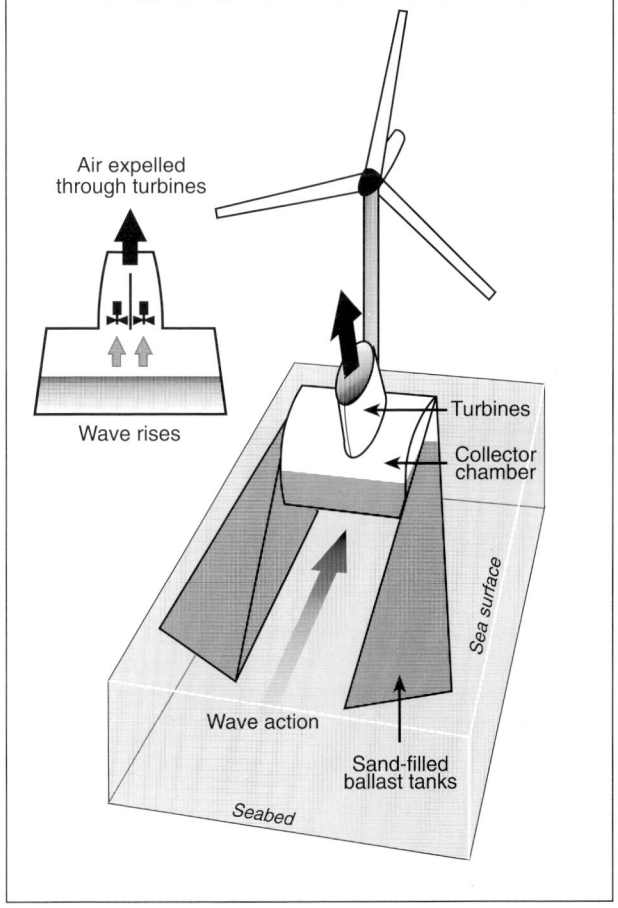

Air expelled through turbines

Wave rises

Turbines

Collector chamber

Sea surface

Wave action

Sand-filled ballast tanks

Seabed

As yet, the technology to achieve wave power generation is undeveloped, and the cost of transmission of energy from our west coasts to centres of population would be high. It is possible that early next century the technology will become commercially available.

Ocean thermal energy makes use of the temperature difference between warm water at the ocean surface and cold water at a depth of 1000 m or more. Although the technique was first proposed in 1881 and a test plant was built near Cuba in the 1930s, little further work has been done until recently. In 1979 a 50 kW test plant was operated off Hawaii, and the Tokyo Electric Power Company has built a plant with a net output of 15 kW in the Republic of Nauru. Again, the technology is in the development stage and capital costs in excess of $20 000 $(kW)^{-1}$ would be expected.

At the present, although New Zealand is well positioned to take advantage of any technology which extracts energy from the oceans, these technologies require considerable development before they could make a major impact on our energy supplies. The country has several more immediate options for renewable energy. However it would be unfortunate if, with its record of innovation, New Zealand was not involved in the advancement of these technologies during the twenty-first century.

12 Energy Efficiency in Transport

Until the middle of the nineteenth century, many people travelled no further than about twenty kilometres from their place of birth during the whole of their lives. Up until that time, horses, or similar animals, provided the only available means of land transport other than walking. Unless people were rich, or had some reason to travel (such as to fight in, or avoid a war, or to undertake a pilgrimage), they would spend their entire life in one locality.

The situation began to change in about 1850, when railways were constructed. This increased the mobility of whole populations, and cities expanded. Although railways made possible the movement of people from location to location, personal transport was still limited to the distance that a person could walk, or if wealthy, ride a horse, in a reasonable space of time. At the end of the century two inventions, the bicycle and the internal combustion engine, changed this situation dramatically.

Although cars began to appear on the roads at the turn of the century, only the wealthy could manage to possess them; this situation did not change until the 1920s, when mass production techniques made family cars affordable. Right through until the 1960s, most people in New Zealand went to work by bicycle, bus, train or by simply walking, unlike today, when fewer than 10 per cent of New Zealand's commuters use public transport, and most people go to work by car.

In the last thirty years, the ownership of cars has increased to the point where there are 1.6 million cars in New Zealand, which is one car for every two people; this gives us the second highest proportion of car ownership in the world. The United States, with a total of 145 million cars, or one car for every 1.7 persons, has the highest rate. In contrast, China – a country with two million cars – has only one car for every 680 people and so a low proportion of car owners.

Personal transport costs not only represent about one fifth of the average New Zealand family budget (see p.30, Fig. 3.4), but petrol consumption accounts for over a fifth of our national energy demand, and represents close to one-third of the energy used by the average household. In the year to March 1994, out of a total consumer energy demand of 410 PJ, the energy supplied by petrol was 94.4 PJ. This represents the use of two million tonnes of fuel each year. The CO_2 produced by New Zealand's motor cars amounts to 40 per cent of total emissions.

Even if we had only to consider the energy requirement of personal transport, then the car would be a major topic, but now it is clear that the motor car is also damaging our atmosphere, cities and personal health, making its impact even more of an issue. Worldwide, motor cars are believed to cause 14 per cent of all CO_2 emissions coming from the use of fossil fuels. This contribution would rise considerably if developing countries such as China came to have similar car ownership figures to ourselves. There is, of course, no reason why they should not have the same freedom of movement as we enjoy. However it is unlikely that China, with a population of over one billion, will ever approach a figure of one car for every two persons, for this would require more than a doubling of the number of cars on our planet. We must realise that New Zealanders are privileged in this respect.

Motor cars put considerable strains on cities. This is not obvious in New Zealand, where our cities have grown on the North American pattern, with adequate room for the motor car, but many cities in Asia and Europe are now having to control motor cars both because of their noise and the atmospheric pollution that they cause. Athens has banned cars from its centre, in order to reduce damage to its ancient monuments and buildings, which are being corroded by atmospheric pollution. Hong Kong and Singapore have moved to limit motorists' access. Hong Kong has experimented with electronic road-pricing, where detectors, set in the road, sense the vehicles passing over them and relay the information to a central computer. Motorists are billed monthly for their road use. In Singapore, access to the central area is controlled by a cordon, which motorists pay a fee to cross. In both cities public transport is cheap, convenient, and widely available. Many European cities have provided large car-parks at the edge of the city, with a cheap shuttle service taking people to their eventual destinations. As cars, moving slowly in congested conditions, are not fuel-efficient, reducing the access of the private motorist to cities will significantly improve overall fuel consumption. Even in New Zealand, if bypass roads were to be built around our major cities, the gains in fuel savings might soon repay the energy used to construct them.

There is mounting evidence that motor vehicles are damaging the health of city-dwellers. Recent observations indicate that vehicles, in particular those which have diesel engines, are a major source of PM 10, which are particles with diameters of less than 10 micrometres. These particles lodge in peoples' lungs; it is estimated that over 10 000 people die each year from PM 10 inhalation in the United Kingdom, a figure which is comparable to the country's road toll. Although no figures exist for New Zealand, it is probable that with our cleaner air and less congested cities the figure is small. However, it cannot be discounted.

It may be that we will find it necessary to reduce the fossil fuels consumed in private and commercial transport. What means have we at our disposal to do this?

Considering the private motorist first, in 1990 it was calculated that the 'average' New Zealander's car had a fuel consumption of 8.7 L /100 km. The 1986 census revealed that the average journey to work for Aucklanders is 8 km, or 16 km return. This is probably higher than the national average. There are 1.45 million employed workers in the country, and it might be guessed that about half drive their cars to work, which indicates that New Zealand workers drive 11.6 million kilometres, and consume 1.3 million litres of fuel daily. A 10 per cent improvement in fuel economy would save over 100 000 litres of fuel every day.

Fuel economy can come from car design, or from better driving techniques. Certainly, cars which have a very low fuel consumption exist. In 1975 Volvo produced a prototype, four-seat car which had a fuel consumption of 2 L/100 km. Recently, the Honda Civic VEi fitted with the VTEC-E engine entered production. In 1992, during an economy rally, one of these cars averaged 2.375 L/100 km or 99 mpg, in the old imperial units. Apart from its fuel economy, this is a normal family car; it makes use of a high gearing ratio, as well as lean burn, electronic fuel injection. If New Zealanders drove similar, fuel-efficient cars, we would more than halve our petrol consumption, a saving of 10 billion litres of petrol each year or of about two million tons of the 7 million tons of carbon we contribute to the atmosphere annually. We have made progress, as the average fuel consumption of new cars purchased in New Zealand fell from 9.7 L/100 km to 8.7 L /100 km between the years 1983 to 1990. Figures supplied by BMW show the trend towards fuel efficiency for their models (Fig. 12.1). Even so, the record for fuel efficiency is an extraordinary 2687 km/L (7591 mpg) set by a team from a French school during a race called the Mileage Marathon, organised annually by Shell. It is unlikely that a practical vehicle could come near to this figure, but the Rocky Mountain Institute (in the United States) hopes to design a car

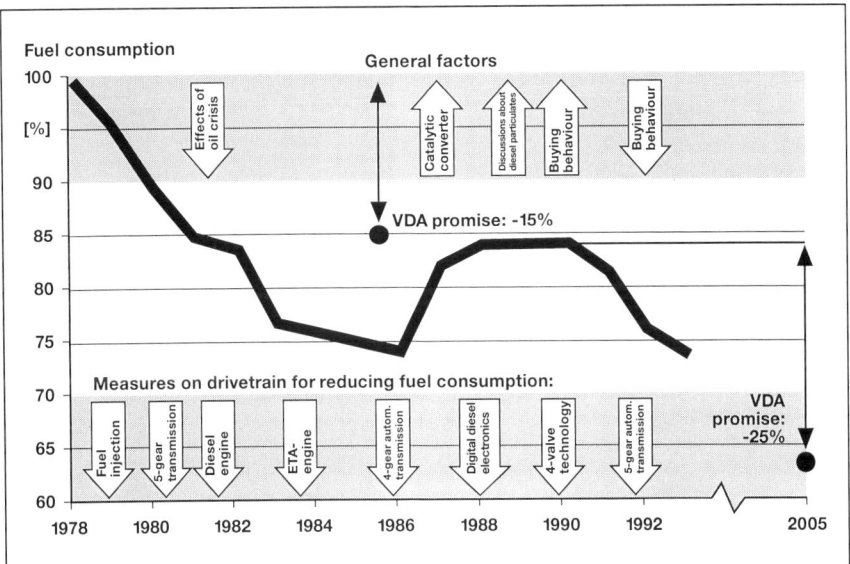

Figure 12.1
Development of
BMW fleet fuel
consumption in
Germany
(weighted by sales
figures).

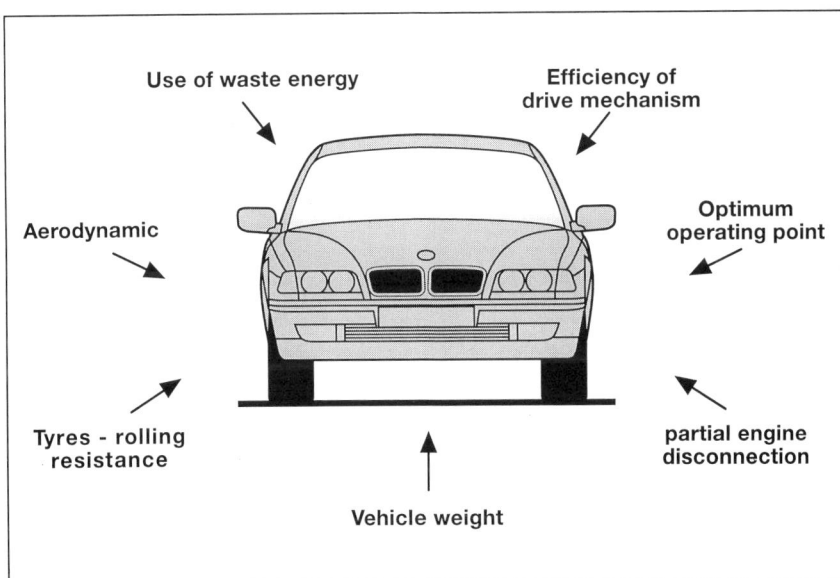

Figure 12.2
Factors in
reducing fuel
consumption.

which will travel 106 km on a litre of fuel (300 mpg). This car will be constructed of modern materials which will reduce its mass to about 600 kg, will use the latest radial tyres which waste about one-quarter of the energy wasted by the cross-ply tyres of the 1970s, and will have a smooth underside, reducing the drag coefficient to 0.14, in contrast to today's average of 0.33. Fig. 12.2 illustrates BMW's strategies for reducing fuel consumption.

Switching to a modern, fuel-efficient fleet would seem to offer considerable savings in petrol consumption, and a reduction in CO_2 emissions. Against this, it must be remembered that some of the energy used in the life of a car, and therefore some of the CO_2 emissions, are due to the manufacture of the car. Because of this, cars which have a long useful life must be considered beneficial. If car manufacturers were to make greater use of recyclable materials, this would have an effect on the amount of energy that a motor vehicle used in its lifetime. Unfortunately, car manufacturers wish to encourage the sale of new cars by 'planned obsolescence', and a continual upgrading of the nation's car fleet means that cars are becoming safer, as new safety features are introduced, and 'old bombs' are scrapped. Some manufacturers such as Volvo make a sales feature of the longevity of their products, and others such as BMW are designing their cars to allow for the recycling of materials. A greater effort to recycle will eliminate the problems of disposal of old car shells.

A related problem is posed by the disposal of scrap tyres. Most of the material, and hence the energy used to produce a tyre, remains when the tread of a tyre has been worn away. Remoulding a tyre is a partial solution, but most motorists do not have confidence in the safety of retreads. In the past old tyres have been dumped, but this represents an environmental problem; if such a dump were to catch fire, as has occurred, then it is virtually impossible to extinguish it, and it can burn for years, emitting dense black smoke. One solution to the problem is to construct power stations to burn these old tyres. Such power stations are now in operation in several parts of the world. Careful design ensures that they emit no noxious fumes, and the steel remaining after the combustion of the tyre can be recycled. An alternative being researched is to heat the old tyres, converting the rubber into a liquid fuel, and recovering the steel reinforcing as before.

Changes in driver habits can make up to a 25 per cent difference in fuel economy. As Fig. 12.3 shows, a car travelling at 130 kmh^{-1} uses twice as much fuel per kilometre as does a car travelling at 50 kmh^{-1}. The graph also shows that steady driving is best for fuel economy.

The main cause of the poor fuel efficiency at higher speeds is aerodynamic drag, which increases as the square of the speed (Fig. 12.4). However, some energy is wasted as tyres flex, especially at high speed. At 100 kmh^{-1} a car wheel does about 20 revolutions each second. This wasted energy appears as heat in the tyres. Both Michelin and Goodyear have developed tyres which contain silica. These tyres waste 20 per cent less energy, resulting in around a 5 per cent improvement in fuel consumption.

Small, light cars are more fuel-efficient than their larger counterparts. There is a catch here, in that (as Fig.12.5 shows) larger cars are safer in accidents. However, as speed is a much more significant factor in accidents, it is possible that we may move to lighter cars and lower, more rigidly enforced speed limits.

Diesel-engined cars make better use of fossil fuels than do those with petrol en-

*Figure 12. 3
Fuel consumption versus car speed.
(Ministry of Commerce)*

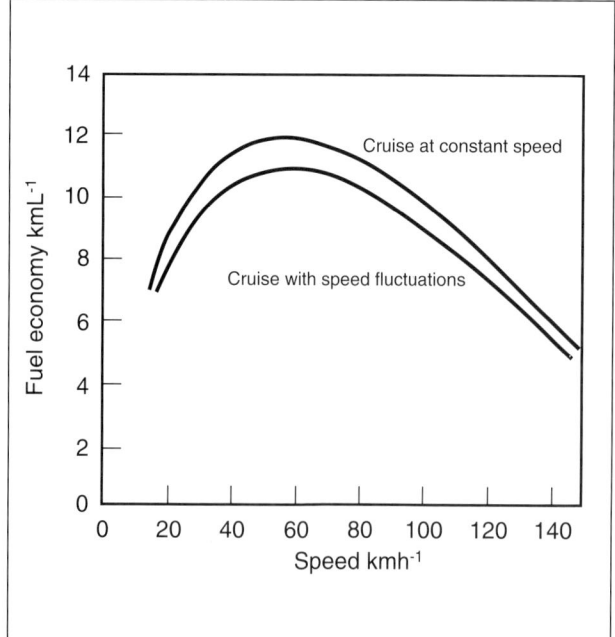

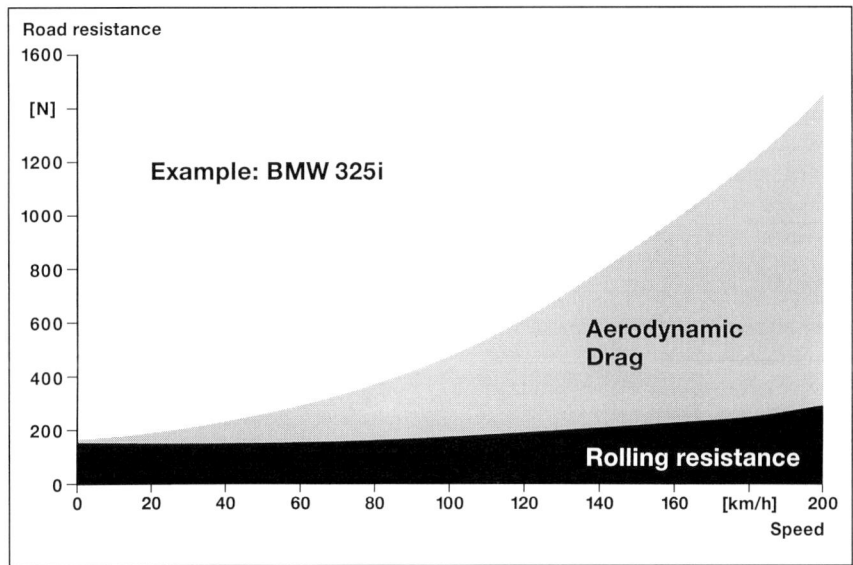

Example: BMW 325i

Aerodynamic Drag

Rolling resistance

Figure 12.4
Aerodynamic drag
and rolling
resistance at
constant speed.

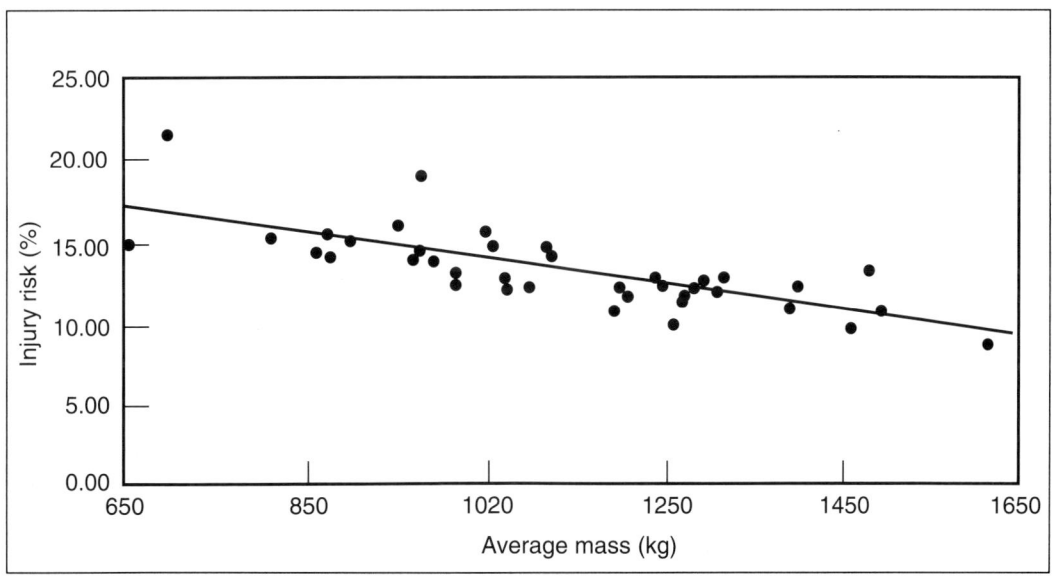

Figure 12 .5
Injury risk versus
mass of car.
(Ministry of
Commerce)

gines. The higher compression ratio of the diesel engine means that it is more fuel-efficient than a petrol engine of the same size. As well, the refining of diesel fuel requires less energy than the refining of petrol; in other words, we get more diesel fuel from each barrel of oil. The effect of this is that if we all switched to diesel-engined cars, our crude oil requirements would be less. Against this, diesel engines are the main source of PM 10 emissions, and so switching to a diesel-engined fleet could damage the health of New Zealanders.

Worldwide, more than 800 000 private cars run on compressed natural gas (CNG), mainly in Russia, North America, Argentina and Italy. The replacement of petrol by this fuel would aid New Zealand for several reasons. We have reasonable reserves of the gas in underground deposits such as the Maui field, and it may be produced from biomass in either digesters or landfill sites. CNG produces only four-fifths the amount of carbon dioxide when it is burnt as does the same mass of petrol, and so use of CNG would reduce our CO_2 emissions. As unburnt methane would disperse rapidly into the atmosphere, the replacement of petrol by CNG would reduce urban pollution caused by unburnt hydrocarbons, providing a good alternative to the fitting of catalytic converters onto our private road fleet. The country has considerable experience in the use of CNG in vehicles, and a distribution network exists in the North Island. Finally, the conversion of a significant proportion of our road fleet could be achieved within a short time.

CNG does have some disadvantages, in that it contains only about half of the energy of petrol for each volume, and so CNG tanks take up much more space in a car, especially as they have to withstand considerable pressure. A further problem is that CNG -fuelled cars suffer a loss of power of about 15 per cent, although this is quite acceptable to most drivers. CNG may also be used as a fuel in diesel engines; Solid State in Lower Hutt, and also a team at Auckland University, have developed designs for this use.

Cars are at their least efficient when the engine is cold. The average car journey made by New Zealanders is about five to six kilometres. Making short journeys like this will halve the fuel economy of a car, so that a car which might use 10 L/100 km on a long trip might use 20 L/100 km around town. A team at Industrial Research, Lower Hutt, have developed a simple device which stores hot water in an insulated container during a trip, and automatically preheats the engine coolant water when the next trip starts. The device can improve fuel efficiency by between 8 and 40 per cent.

The realisation that the car is very inefficient over short journeys suggests that it is not a good means of movement within cities, where average journeys may be as short as one kilometre. Although this trend has not yet come to New Zealand, many cities are encouraging the use of the bicycle, which is a highly efficient form of personal transport. In Seattle the police have used bicycles for several years, and have found that within the inner city, police cyclists are more mobile, and better able to move rapidly to where required. Many big cities have bicycle courier organisations, where the speed and manoeuvrability of cyclists in dense traffic, and the ease with which a bicycle can be parked, have given them an advantage over the motorist. The bicycle is the preferred mode of personal transportation in several cities in Northern Europe, in particular those in Holland, as well as in many university cities around the world.

If bicycles are to regain a greater share of personal transportation, then provision must be made for them in the form of cycle lanes, for cyclists are at great risk when occupying the same road space as cars, buses and lorries. As well, adequate storage space has to be provided for bicycles. In some US cities, employers are required to provide not only bicycle storage, but suitable changing facilities for cyclists as well.

Better public transport should be encouraged within cities. This will require innovative thinking on the part of city councils, who will have to be prepared to subsidise transport companies. Several experiments are already underway in the congested European cities, particularly those which have made bus-only lanes, in which buses can move unrestrictedly during peak travel periods. Door-to-door minibus services, which are similar to those operated to and from airports at the present time, are already filling the gap between expensive but flexible taxis and cheap but inflexible buses. Such shuttles

City circuit electric bus, Oxford (UK). (Photo: Southern Electric, Berkshire)

reduce urban traffic. Further into the future, the return of electrically driven buses, either powered by internal batteries or by overhead power cables, or of buses which store energy in flywheels, may provide non-polluting urban transport. Oxford (in the United Kingdom), a city in which pollution levels can rise well above permissible levels, introduced four battery-powered buses into service in late 1993. These 18-seat buses are powered by lead-acid batteries which can store 230 MJ (64 kWh) of electricity. The buses require fifteen minutes' recharging every hour, as well as eleven hours of charging overnight. A flywheel-driven bus is in use in Himley (UK). This bus is powered by a one-metre diameter flywheel, which weighs 250 kg. The flywheel is accelerated to operating speed in 90 seconds, using electrical energy. This can be done at stops provided with electrical outlets, allowing the bus to operate regular services all day. Manchester (UK) has a 'high tech' tram which runs at low speeds on city roads and then converts to high speeds on railway track outside of the city centre.

The movement of freight is essential to New Zealand's continuing prosperity. Heavy goods vehicles and buses use about 1.3 million tonnes of fuel each year, and fuel amounts to about 15 per cent of the operating costs of a truck. Savings can be made by limiting the speed of trucks. An on-board computer can be fitted to a truck to control its speed, with a saving in fuel, maintenance and tyre-wear. For a truck with a fuel consumption of 0.5 L/km, the saving can easily be $2000 a year, which means that the cost of the computer ($1800-4000) is rapidly paid back. Greater use of the rail system for freight would save fuel and CO_2 emissions, especially if further electrification of line occurred. Since the North Island main trunk line between Palmerston North and Hamilton was electrified, 20 million litres of fuel has been saved each year. The 1700 kW DX diesel locomotives were replaced by 3000 kW electric locomotives operating at 25 kV. These locomotives have regenerative braking, which converts kinetic energy back into electrical energy; this saves 13 per cent of the energy required. Suggestions have been made that electrified rail lines might use energy generated by wind turbines, sited along the railway line at places with good wind resource. The excess energy generated could be transmitted by the overhead cables, and sold into the national grid.

Further savings would occur if freight were to be carried by sea. Most of New Zealand's cities are on the coast, or have easy access to ports. The movement of containerised freight by sea is efficient in energy terms, as not only can marine diesel engines achieve 50 per cent efficiencies in the conversion of fossil-fuel energy to motive energy, but a ship can carry several thousand tonnes of freight. A greater movement of freight by sea would also

Walker Wingsail trimaran, Zefyr 001, photographed in November 1994. (Photo: Walker Wingsail Ltd)

remove some of the large trucks from our roads and city streets, with a corresponding reduction in the congestion of and damage to roads, and improvement in atmospheric and noise pollution in cities. It may even be possible to use wind energy to assist ships' propulsion. During the 1920s two German ships, the *Buckau* and the *Barbara*, were equipped with Flettner rotors, devices which act as sails. The *Buckau*, renamed the *Baden-Baden*, crossed the Atlantic in 1926. Little data survives from these experiments, but it seems that both ships could gain up to one-third of their motive force from these rotors. In 1980, a small Japanese tanker, the *Shinaitoku Maru*, and a 3000-tonne American freighter, the *Mini Lace,* were fitted with square sails of modern design, and in 1986 the 6500 tonne *MV Ashington* was fitted with three 'wingsails', a development pioneered by Walker Wingsail of Plymouth (UK). This firm has developed high-tech trimarans which have proved capable of operation in all conditions. In all cases fuel and pollution savings greater than 10 per cent were obtained, and in the case of the *Ashington* these savings reached 36 per cent, with six-and-a-half hours cut off the journey time during a North Sea crossing. The experiments were discontinued because low oil prices made fuel economy less important.

Since 1980, Norwegian researchers have been developing foils which fit on the bows of ships and extract energy from waves. A 20 m research vessel has managed to extract 15-22 per cent of the energy needed for its propulsion during tests in 3 m high waves. As ocean waves typically contain 60 kWm^{-1}, a 70 m ship with waves abeam might be passed by 4200 kW. As such a ship would need 1000 kW to maintain 18 kmh^{-1}, extraction of some of this energy could result in considerable fuel savings.

Marine engineering is a conservative technology. It took sixty years for the screw propeller to become accepted. Nev-

ertheless as we have recently proved at San Diego, New Zealanders are innovative sailors. If we made fuel savings in our maritime fleet, it would aid our country. Furthermore, as the world's shipping fleet consumes fuel equivalent to 730 million barrels of oil each year, if 10 per cent savings could be achieved worldwide, this would save 73 million barrels of oil, the CO_2 emissions caused by the burning of this oil, and the $3 billion cost of the oil.

It is clear that considerable savings of fossil fuels and CO_2 emissions could be made in the transport sector. However, this sector is one in which many commercial factors are paramount, and so savings are likely to come through the operation of 'market forces', unless environmental concerns are translated into legislation. The introduction of carbon taxes could be a first step towards improving the environmental impact of the transport industry in our country.

Car manufacturers have begun to develop electric cars. As discussed in the earlier section on the storage of electrical energy, one problem hindering the intro-

duction of an electric car has been the lack of a good battery system. Even so, a number of manufacturers have produced prototypes. Peugeot propose to begin production of the 106 EV in 1995. This car, based on the Peugeot 106, has a range of 120 km at 80 kmh[-1], which is more than adequate for commuters. It stores energy in a nickel-cadmium battery pack, and has an on-board charger in the boot.

California has passed a zero-emission law requiring manufacturers to ensure that at least two per cent of the cars they sell in 1998 are zero-emission cars. This figure must rise to 10 per cent by 2003. The states of Massachusetts and New York now have similar laws, and other states are proposing to follow suit. This has forced American car manufacturers to design electrically powered cars and commercial vehicles. Ford has developed the Ecostar, based on its European Escort van. This van uses state-of-the-art sodium-sulphur batteries, giving it an 120 kmh[-1] top speed and a 400 kg payload. Chrysler have modified their T-Van. Using nickel-iron batteries, this van has a 100 kmh[-1] top speed and a range of

The Peugeot 106 EV, an electric car planned for production in 1995. (Photo: Peugeot Concessionaries, NZ)

130 km. The van has been in commercial production since April 1993. General Motors have developed the Impact, in an attempt to design a practical electric car which meets all safety standards, and which customers will wish to buy. This car, which stores energy in lead-acid batteries, has a 100 kW motor, a kerb weight of 1300 kg and a drag coefficient of 0.19, giving it 'normal car' performance. Although its range is only 110 km in cities, extending to 160 km with steady motorway driving, it can be recharged in two hours. GM expect to begin sales in 1998.

Other European and Japanese car firms have produced concept designs, but as yet have no production plans. Initially, electric cars will cost more than their petrol-engined counterparts, although maintenance will be cheaper, even allowing for the need to replace batteries every few years. Research indicates that these cars will be bought initially by affluent people who either support 'green' movements, or enjoy possessing new technology.

If the problem of the electric car's range cannot be overcome, 'hybrid' cars may help to reduce emission problems. Hybrid cars have small liquid fuel motors which continuously charge a battery pack. The car is then driven by electric motors in the same way as an all-electric car. Hybrid systems could double fuel-efficiency and reduce emissions dramatically. The traditional car has an engine which develops powers in the range of 50-100 kW to ensure adequate acceleration. When motorway cruising, it requires only a sixth of this power and in cities even less. This means that the engine is running inefficiently for most of the time. The hybrid car will have a small motor, possibly developing only 5-10 kW. This engine is run constantly at its most efficient speed, charging the battery pack, which can be much smaller and lighter than in an all-electric car. If the pack is fully charged, the charging motor is shut down. The car can use regenerative braking, thereby returning kinetic energy to store. In cities, the recharging motor can be switched off and the car driven as an all-electric car, completely eliminating pollution. The charging motor, which might even be a fuel cell, could be fuelled by methanol or ethanol derived from biomass, and so the hybrid would not require fossil fuels for its energy source.

If we consider the physics of a typical production car driving at 60 kmh⁻¹, we find that 80 per cent of fuel energy is lost to heat in the engine and gearbox, and so at best only 20 per cent of the energy of the fuel is available for motion. Of this motion energy, about half is lost to air resistance, and half to rolling resistance of the tyres. Further energy is lost when the brakes heat as the car slows. All of these losses can be reduced. Our hybrid would have an engine which has been designed to run at maximum efficiency and would have no gearbox. Although some losses will occur in converting the mechanical energy into electrical energy and back again, we might expect that more of the fuel energy is available for maintaining motion.

Because the hybrid can have a light-

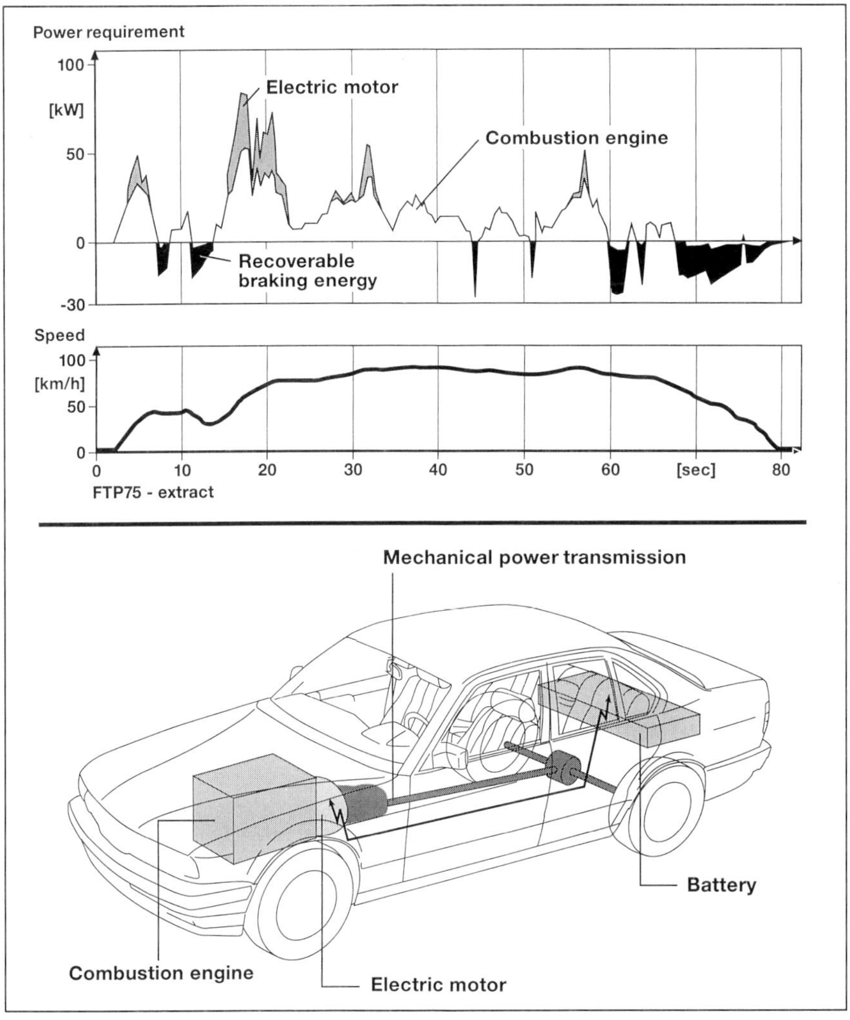

Figure 12.6
Parallel hybrid drive. Top: diagram showing operating strategy. Bottom: design of car.

weight motor, the car itself can be highly aerodynamic and of low mass, halving air and rolling resistance. Coupled with the regenerative braking, which could recover virtually all of the energy loss due to braking, the motion energy losses could be reduced to one third of those at present. This would have a dramatic effect on fuel consumption, because if four-fifths of the energy is lost in the engine and gearbox region, every megajoule saved in motion energy really saves 5 MJ of fuel energy. It has been calculated that such a hybrid car could achieve over 100 km/L, with the recharging motor being used continuously, and if battery power alone were used for the short trips, this figure could be doubled.

BMW have made extensive studies of the hybrid car concept, and consider that such a car could be commercially produced within the immediate future (see Fig. 12.6).

That the technology to produce ultra-efficient personal transport exists has been demonstrated in the World Solar Challenge race held from Darwin to Adelaide. In the 1994 event, the winning car, the Honda 'Dream', averaged 85 kmh[-1] for the 3000 km race and used so little energy that the total cost of this electricity would have been $5 if purchased from the mains. The car used solar cells which were developed by the University of New South Wales and had 20 per cent efficiency, giving a maximum power of 1.6 kW. The car had electric motors which were fitted inside the hubs of the wheels and computer control gave a gearing effect. The drive had an overall efficiency of 97 per cent. New tyres reduced rolling resistance by 30 per cent and put them on a par with the best cycle-racing tyres. High technology batteries were also used. Although these vehicles do not seem practical at present, it must be remembered that it is unlikely that one

of the first 'horseless carriages' produced in 1894 could drive from Darwin to Adelaide, let alone at 80 kmh[-1].

It is likely that we will see the production of either all-electric or hybrid cars which gain some of their energy from solar cells within the next twenty years. As our present cars use the energy contained in fossil fuels with at-the-most 20 per cent efficiency, after the petrol has been refined, it could still be an advance to generate electrical energy in fossil-fuelled power stations at 40 per cent efficiency and use this energy in electric cars which might be 60 per cent efficient, if we include transmission losses in the electrical grid in this figure. This would give an overall efficiency of 24 per cent, based on the energy contained in the fossil fuel, as well as having the beneficial effect of removing pollution from cities. Even so, this would not represent a great advance, unless there were a move away from fossil-fuelled power stations towards those deriving their energy from renewable forms.

Another consideration is that the future may see a reduction in the amount of work-related travel. Advances in electronic data transfer, the so-called 'information superhighway', mean that it is now possible for some employees to work from home, or to avoid a lot of business travel. Teleconferencing in particular will allow a considerable reduction in domestic and international travel.

Finally, there is often the alternative of living so close to one's work that it is usually practicable to walk to work. This is the choice of one of the authors, who walks to his office most mornings through one of the most interesting and attractive city Botanic Gardens in the world! The other author appreciates his harbourside views so strongly that he lives twenty kilometres from the city. Nevertheless, he uses public transport.

Solar Kiwi Information

Weight	200 kg
Height	1.1 m
Length	4.5 m
Top Speed	100 kph
Cruising	60-65 kph
Body	Carbon fibre / Kevlar / Divincell composite
Wing	Nomex / Carbon
Brakes	Hydraulic drum front wheels Regenerative rear wheel
Suspension	Rear: Trailing arm Front: Leading beam with Lurothane springing
Solar Array	Poly-crystalline 8 sq m
Motor and Controller	96V Brushless permanent magnet
Batteries	Sealed lead-acid 96V system 1.8 kwh
Wheels	Kevlar carbon composite discs
Tyres and Rims	Panaracer
Suspension Components	Chrome-moly
Transmission	Lightweight high efficiency chain drive
Wing Tilt Mechanism	Driven by linear acutuator

'Solar Kiwi' car, sponsored by Waikato Electricity, which performed well in the 1993 World Solar Challenge race from Darwin to Adelaide. It was first in the Private Entry Class, for Using Lead-Acid Battery, and for Using Commercial Solar Cells. (Photo: Waikato Electricity)

13 Strategies for the Future

We would be foolish to claim that we can predict the future, particularly in the case of energy use and its political, economic and environmental impact. This point can be well illustrated by considering the predictions that have been made for trends in oil prices over the years 1968-93 (Fig. 13.1).

A further example of a prediction which was wide of the mark, is found in the 'Club of Rome' warning, made in the 1960s, of global famine by the 1990s. This famine did not occur because although the world's population rose as expected, scientific advances allowed global food production to rise at a similar rate. This is an example where a 'technical fix' has, to date, averted a predicted disaster.

Uncertainties are always inherent in politics and economics where the decisions of individuals, some of whom do not necessarily behave rationally, can affect the course of world events. Even when trends, such as the increase in the concentration of CO_2 in our atmosphere, can be predicted with confidence, the likely effect of these trends may be unclear. Some groups have suggested that global warming may not be an inevitable consequence of increased CO_2 levels, and factors such as changes in the Sun's radiation might be affecting our planet. Even today, the evidence that global warming is proceeding is not clear-cut; most climatologists believe that it will be another decade before sufficient data can be assembled to show that the present warming trend is more than a short-term fluctuation.

Having granted these uncertainties, we have already received warnings in the forms of acid rain, ozone depletion, and rising sea levels. In the first two of these cases, large-scale damage has been done to our environment by substances which we have released into the atmosphere, and the economic cost of this damage is certainly such that it outweighs the short-term gains made by allowing these substances to be released in the first place. In both cases, governments, some of which are not noted for their environmental concern, have been forced to take action to protect their environment, and that of other countries. For instance, the combined effect of

Figure 13.1
Trends in oil prices, 1968-1993. (World Renewable Energy Congress 1994)

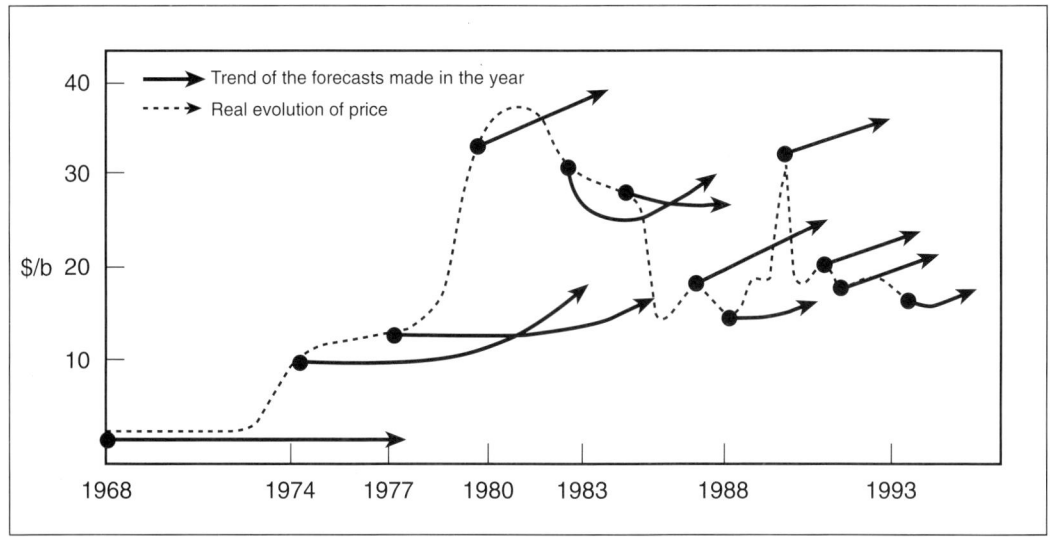

the sea levels rising by 12 cm and the land sinking by an equal amount has posed a considerable threat to the city of Venice. This threat to one of the world's most historic cities was reported in the February 1995 *National Geographic* magazine and is seen as an international tragedy that must be averted.

Most climatologists, as represented by the Inter-Governmental Panel on Climate Change (IPCC), are convinced that there is a high probability that global warming is proceeding, and that this is caused by the increasing concentrations of greenhouse gases in our atmosphere. Governments of over 160 countries met at Rio de Janeiro in 1992 and signed the Climate Convention which was ratified on 21 March 1994. This convention recognises that changes to the Earth's climate are of common concern to humanity and that increasing concentrations of greenhouse gases 'will result on average in an additional warming of the Earth's surface and atmosphere and may adversely affect natural ecosystems and humankind.' The convention states governments 'should take precautionary measures to anticipate prevent or minimise the causes of climate change and mitigate its adverse effects'. In this respect, the governments are taking a risk-averse approach, accepting that the risk of climate change is serious, and that measures have to be taken now, as it will be too late to avert disaster if they wait until the evidence is conclusive.

If the world continues with 'business as usual', levels of CO_2 in the atmosphere will double by 2100. Even if CO_2 emissions are held at the 1990 levels, concentrations will continue to rise for several centuries before stabilising at a level over twice that of today. Only by cutting emissions by 60 per cent can we stop the build-up of CO_2. The global temperatures graph in Chapter 2 (Fig. 2.6 on page 23) shows how temperatures are expected to rise if emissions are not controlled.

The developed nations, a group to which New Zealand belongs, are the major producers of CO_2. It is probable that the United States contributes 25 per cent of global emissions, at the time of writing, while closer to home, Australia has as great an impact as does India, a country with almost fifty times Australia's population. We cannot expect the developing nations to pay attention to global warming if the developed nations do not pay equal concern.

How then can we in New Zealand, a small country, play our part? In March 1994 we ratified the Climate Convention, and so we should be developing methods in order to stabilise or reduce our net greenhouse emissions. New Zealand has a reputation for social innovation, and with our 'clean green' image we could gain international benefits from taking a lead, doubtless enhancing our tourist trade as a consequence. Increased tourism, combined with economic advantage from better use of energy, would benefit our country.

There are difficulties in estimating our net greenhouse emissions. Methane produced by farm animals and nitrous oxide evolved by nitrogenous fertilisers add to our impact on the atmosphere, and these cannot at present be accurately assessed. However our contribution to atmospheric CO_2, which is increased when we burn fossil fuels and remove trees and bush, and is reduced when we plant trees and bush, can be estimated and monitored. In the struggle to control CO_2 emissions, New Zealand is lucky, for we are likely to increase our tree planting into the foreseeable future, as forestry is a major industry. This will reduce our net CO_2 emissions, even if we continue to burn fossil fuels at our present rate.

It is unlikely that humankind can rely on 'market forces' to aid our attempt to control global warming. Market forces are reactive, and rely on the financial outcomes of actions to determine the directions which economies should take. Environmental costs do not affect financial outcomes unless carbon taxes or regulations are put into place, and few countries are at present contemplating such measures. Furthermore, market forces take little or no account of the social costs of economic policies.

Some economists have attempted cost-benefit analyses of measures to slow global warming. Generally they assume that a doubling of CO_2 levels will only lead to damage in the order of 2 per cent of the world's GNP. These economists conclude that as it would cost over 2 per cent of GNP to stabilise CO_2 emissions, the task is not worthwhile. As most climatologists put the probable cost of damage far higher, it seems risky to wait until it is certain that global warming is occurring, assuming that countries will eventually act if damage is much higher than 2 per cent of GNP. Certainly it is difficult to tell the inhabitants of Pacific islands that the developed world will not take precautionary actions against their countries being submerged by rising sea levels or destroyed by cyclones because it might cost people living in the developed world a few hundred dollars each, per annum. Equally the owners of seaside property in our country may not be keen to abandon their 'bach' so that the inhabitants of Los Angeles can continue with business as usual. If market forces are to be the final arbiter of energy policy in this country, then consumers will need to use their power in the market to force energy providers towards renewables.

In the past, the planning process in New Zealand has been controlled by small but powerful groups, and this has led to environmentally unsound decisions. We believe that the present move towards proportional representation indicates that voters wish to have greater control over the decision-making process, and will be prepared to force politicians and industry to take steps to protect both our country and our planet.

We suggest that New Zealand could adopt a policy to stabilise CO_2 emissions at 1990 levels, based on the encouragement of renewable energies and of energy conservation, coupled with an increase in tree planting to soak up CO_2. In a country such as ours, the real answer lies in education. A well-informed public, which understands the issues and options, is in a good position to cause political action to occur.

New Zealand is already a world leader, in that it produces 75 per cent of its energy from renewable sources. As we have pointed out, we could generate 10 per cent of our electricity from the wind, using perfectly standard present-day technology, at a cost no higher and with less environmental impact than that of building more hydro dams. In the short term it may be cheaper to build a gas-fired power station in Taranaki, but in the long term this would probably be the most costly option for increasing our electrical generation. Biomass offers New Zealand the opportunity to stabilise CO_2 emissions and at the same time develop exportable expertise in energy production. Donald Rosenthal of the US Dept of Energy has estimated that by planting trees to mop up CO_2, the US could go on with business as usual until 2015. Calculations show that planting a million hectares of coppiced poplar trees could counter a 3 per cent increase in the UK's CO_2 emissions. In this country, the recent recommendation made under the Resource Management Act that should ECNZ proceed with the construction of the 400 MW natural gas-fired power station, then sufficient trees to completely 'offset' the CO_2 emission would have to be planted, is a clear recognition of the potential of tree planting for locking up atmospheric CO_2.

In the United Kingdom, the government has introduced a Non-Fossil Fuel Obligation (NFFO) which gives subsidies to electrical generation by methods not using fossil fuels. The NFFO was designed to make nuclear power sufficiently economic for the nuclear generation industry to be privatised. This did not occur because investors would not accept the hidden costs of decommissioning nuclear power plants. The NFFO has instead given a boost to renewables, and if a similar scheme were to be introduced into this country it might well be all that is required to induce utilities and private investors to make a substantial investment in renewables. The United Kingdom, which has a market-led economy similar to our own, is starting to make us look very backward.

New Zealand is well placed to improve efficiency in the use of energy. Our towns and cities are spacious, our climate is moderate, and as a nation we are prepared to accept new ideas. If in the future we encouraged new building to conform to a high standard of design, we could reduce our energy consumption. As an example of what can be done, the Inland Revenue building in Nottingham (UK), which was completed in 1994, makes use of computers to control natural ventilation. It is expected that the building will use about 360 MJm^{-2} (100 $kWhm^{-2}$) of energy each year, as opposed to a conventional building of the same size which would use four times as much. The Nelson Public Library has been constructed to a similar efficiency, as we have seen.

In the field of transportation, greater use of teleconferencing and electronic mail would reduce costs and greenhouse emissions significantly. New Zealanders have proved to be very innovative in their use of electronic communication, and we would expect that they would welcome the opportunity to avoid tiring early morning flights to business meetings. Within our cities, the provision of better public transport, particularly if electric-powered, would reduce congestion, noise and pollution, and the provision of facilities for cyclists would reduce not only the use of fuels but also the very considerable cost of injuries to cyclists.

It has been suggested that the construction of a national 'superhighway' might save transport fuel. If this highway were to bypass all towns and be provided with recharging facilities for electric cars, then motorists could travel to their destinations without being held up by lorries or in towns. Unfortunately, the experience of the United Kingdom is that the building of motorways simply causes people to buy more cars and so congestion is, if anything, increased. It must also be remembered that motorway building requires captial which could be used to improve the public transport system. Such motorway building would only be useful if it led to a reduction in both fossil fuel use and road accidents. It will be necessary that the New Zealand public understand the issues involved, before decisions on such matters are taken.

We in New Zealand could easily reduce our CO_2 emissions back to our 1990 levels, if not further, by 2000. The technology exists for us to reduce our dependence on fossil fuels considerably, and if the government were to develop a programme to plant trees along our roads, this and forestry planting, and pest control, would reduce our CO_2 emissions further. In this enhancement of our 'clean green' image there would be considerable benefits to tourism. Even to be able to say that we generate 100 per cent of our electricity from renewable sources would be valuable, both to attract the 'ecotourist', and to give us status in international negotiations, particularly with our Pacific island neighbours who are most at risk from global warming, and who will certainly look to us as sea levels rise.

Of course, one absolutely fundamental problem for our world is population control. If we wish to retain our unique quality of life we have to prevent excessive population growth on these islands, and may even, in the future, have to limit the numbers of visitors who come here. Population control is ultimately the only way in which energy consumption can be controlled, and we in New Zealand are in a good position to lead the world in this aspect.

It is our conclusion that New Zealanders must face the probability that our present way of life, and our seeming ability to increase our energy demand indefinitely into the future, is not sustainable. Changes in attitudes will have to be made by both individuals and by industries. Although environmental and social costs and the depletion of fossil fuel reserves require us to reduce our fossil fuel consumption, many see climate change due to CO_2 emissions to be the urgent problem. If we do not achieve change voluntarily, then events might force it upon us in a way which might prove to be less comfortable than otherwise. We hope that in this book we have made a good case for our beliefs. If we have, then the rest is up to all of us!

Glossary

AC

Alternating current which changes magnitude and direction regularly, following a sine function. In New Zealand, the frequency of this sine function is 50 Hz.

Acid rain

Rain usually has a pH of about 5.6, due to its having dissolved some atmospheric carbon dioxide. If the rain has dissolved some sulphuric or nitric acid formed from industrial pollutants, its pH will be less than this value and the rain is termed 'acid rain'.

Biomass

The total mass of living material in a given area. When applied to renewable energy, this refers to fuels derived from organic material, cropped from some area.

Business-as-usual scenario

The scenario which assumes that there will be no change of attitude or energy consumption patterns in the future.

Capacity factor

The ratio of the average power output of a power plant to its rated output.

CFC

Chloroflurocarbons. These are compounds which have been used extensively as refrigerant fluids and as aerosol spray propellants. Not only are they powerful greenhouse gases, but they also damage the Earth's protective ozone layer. Their manufacture and use is being scaled down under the Montreal protocol agreement (1987).

Chemical potential energy

Useful energy contained within a substance by virtue of its chemical composition. This energy may be released to perform useful work by means of suitable chemical reactions, often an oxidation such as combustion.

CO_2 emissions

The release into the atmosphere of the greenhouse gas carbon dioxide, generally caused by the burning of fossil fuels, although extraction of natural gas and geothermal energy can cause these emissions.

Coppicing

The pruning of trees in such a manner that fuel is obtained but the trees are not destroyed, but may regrow to produce more fuel.

(The) cumec

A water flow of one cubic metre per second past any point of measurement.

DC

Direct current which flows in the same direction all of the time. A cell or battery produces DC.

Energy

The ability to do work. A system is said to possess energy if some useful work can be extracted from that system. Energy is measured in the SI unit, the joule.

Energy intensity

The amount of energy required to produce each unit of GDP. It may be measured in $kJ\$^{-1}$ or in arbitrary units, often related to the energy intensity for a particular year.

Enhanced greenhouse effect

Global warming caused by the trapping of radiation, emitted by the Earth, in the atmosphere due to the presence of gases produced by human activity.

Fissile

An isotope is said to be fissile if its nuclei can be split into two smaller nuclei, of approximately equal sizes, either spontaneously, of on absorption of a neutron.

Fission

The process whereby the nuclei of a fissile material split into two approximately equal halves.

Fossil fuels

Fuels such as coal, oil and natural gas, formed from the remains of plant and animal materials.

Global warming

The (suggested) raising of the Earth's surface temperature due to the increasing concentrations of greenhouse gases.

Gravitational potential energy

Energy which a mass possesses by virtue of its position in a gravitational field.

Gross domestic product (GDP)

The total production of a country measured as the monetary value of all (final) goods and services produced.

Gross national product (GNP)

The total income of a country measured as GDP plus share dividends, interest and rents coming into that country from abroad, minus share dividends interest and rent flowing outwards.

Head (of water)

The height difference between the water level at the intake of a hydro power station and at the outflow.

Intercalation

The storage of small ions, such as the lithium or hydrogen ion, within the lattice of an electrode.

(The) joule

The work done when a force of one newton moves through one metre, in the direction of that force.

(The) kilowatt-hour

The energy transferred when a device with a power of one kilowatt operates for one hour. It is equivalent to 3.6 megajoule (3.6 MJ).

Kinetic energy

The energy possessed by an object by virtue of its motion. The value of the kinetic energy of a body is expressed by the relationship $E = \frac{1}{2}mv^2$

Load factor

For a power plant, this is the ratio of the average power output to its peak power output, usually calculated over a one-hour period.

Natural greenhouse effect

Even in the absence of human activity, the Earth's atmosphere contains gases which trap outgoing radiation, thereby raising the average surface temperature by about $20°$ C. This is the natural greenhouse effect.

Payback time

The time it requires for the financial savings obtained through some energy-saving technique to equal the cost of installation of that technique.

Per capita energy consumption

This is calculated as the ratio of the annual energy consumption of a country to its population.

Photochemical smog

A smog of droplets of organic chemicals formed by reaction between unburnt hydrocarbons emitted by motor vehicles, and oxides of nitrogen and ozone. These smogs are often formed above cities such as Los Angeles and Mexico City, which are surrounded by hills which shelter the cities from winds.

Photosynthesis

The series of reactions by means of which plants take up carbon dioxide and water and in the presence of sunlight convert these into materials for their growth, giving off oxygen as a by-product.

Photovoltaic cell
A single device for converting some of the energy of sunlight into electrical energy. A group of such cells may be linked to form a module.

Potential difference
The difference in electrical energy, measured in volts (V), between two points in an electrical circuit. A potential difference of one volt between two points in a circuit means that if one coulomb of charge were to flow between those two points, then one joule of energy would be transferred.

Power
The rate of doing work; the rate at which energy is transformed from one form into another. The SI unit of power is the watt (W), which represents an energy transfer of one joule per second ($J\ s^{-1}$).

Primary energy
Energy sources, for example wind or nuclear, which are not used directly, but are converted into other energy forms such as electricity for use.

Rated power
The design power of a machine, or power plant.

Renewable energy
Energy which does not require the depletion of a pre-existing source or supply for its production.

Resistance
This is the ratio of the potential difference across a circuit component to the current flowing through it. The SI unit of resistance is the ohm (Ω).

Resistive heating
The production of heat energy by the direct conversion of electrical energy into heat energy in a resistor. Most domestic electrical heaters are of this type.

SI
The letters SI stand for Système Internationale, a system of units based on careful definition of the metre, second and kilogram, as well as some other basic units. Further units are derived from these.

Sustainable energy policy
A sustainable energy policy would be one by which a society meets its present energy needs without compromising the ability of future generations to meet their needs.

Thermal equilibrium
A condition in which an object receives and emits the same amount of heat energy. Such an object will maintain a constant temperature.

(The) watt
The SI unit of power. A device is said to be operating at a power of one watt if it converts one joule of energy from one form to another each second.

Bibliography

Dang, Hien D.T. (1994). *Energy Data File*, Wellington: Ministry of Commerce

Harris, G. et al (1993). *Promoting the Market for Energy Efficiency*, Wellington: Ministry of Commerce (Energy and Resources Division)

Houghton, John (1995). *Global Warming: The Complete Briefing*, London: Lion Press

Renewable Energy Opportunities for New Zealand (1993), Wellington: Ministry of Commerce

Salinger, Jim (1991). *Greenhouse New Zealand*, Dunedin: Square One Press

Sayigh, A.A.M. (ed.) (1994). *Renewable Energy: Climate Change, Energy and the Environment*, Oxford: Pergamon (Proceedings of World Renewable Energy Congress, 11-16 September 1994)

Wind Energy: Guidelines for Renewable Energy Developments (1995), Wellington: Energy Efficiency and Conservation Authority

World Energy Council (1993). *Energy for Tomorrow's World*, London: Kogan Page Ltd

World Energy Council (1994). *New Renewable Energy Resources: A Guide to the Future*, London: Kogan Page Ltd

The authors found *New Scientist* to be an invaluable source of information on current developments in renewable and sustainable energy topics.

Index